OPTIMUM COMPOSITE STRUCTURES

Edited by **Karam Y. Maalawi**

Optimum Composite Structures
http://dx.doi.org/10.5772/intechopen.74696
Edited by Karam Y. Maalawi

Contributors

Rashmi Aradhya, Nijagal M. Renukappa, Karam Youssef Maalawi, Bo Jin, Sergey Golushko, Ferhat Fedghouche, Sergey Shevtsov, Igor Zhilyaev, Natalia Snezhina, Jiing-Kae Wu, Sang Yoon Park, Won Jong Choi, Ruham Pablo Reis, Edson Botelho, Américo Scotti, Alberto Santos, Leonardo Sanches, Iaroslav Skhabovskyi

First published in London, United Kingdom, 2019 by IntechOpen
IntechOpen is the global imprint of INTECHOPEN LIMITED, registered in England and Wales, registration number: 11086078, The Shard, 25th floor, 32 London Bridge Street
London, SE19SG – United Kingdom
Printed in Croatia

British Library Cataloguing-in-Publication Data
A catalogue record for this book is available from the British Library

Additional hard copies can be obtained from orders@intechopen.com

Optimum Composite Structures, Edited by Karam Y. Maalawi
p. cm.
Print ISBN 978-1-78985-067-3
Online ISBN 978-1-78985-068-0

For EU product safety concerns:
IN TECH d.o.o., Prolaz Marije Krucifikse Kozulić 3, 51000 Rijeka, Croatia,
info@intechopen.com or visit our website at intechopen.com.

Meet the editor

Karam Y. Maalawi is a professor of aeronautics and mechanics at the Mechanical Engineering Department, National Research Centre in Cairo, Egypt. He obtained his M.Sc. and Ph.D. degrees in aerospace engineering from Cairo University. Throughout his career, he has utilized the vast range of knowledge at his disposal to contribute to 10 research projects regarding aerospace engineering technology, wind turbine components and renewable energy applications. Likewise, Dr. Maalawi has published extensively in the field of structural optimization and wind turbine design and performance. For his professional accomplishments, the National Research Centre has recognized him with two awards for engineering sciences. Dr. Maalawi is a consultant member of the Egyptian Organization for Standardization and Quality.

Contents

Preface

The subject of optimum composite structures is a rapidly evolving field and intensive research and development has taken place in the last few decades. Therefore, this book aims to provide an up-to-date comprehensive overview of the current status in this field to the research community. The contributing authors combine structural analysis, design and optimization basis of composites with descriptions of the implemented mathematical approaches. Within this framework, each author has dealt with the individual subject as he/she thought appropriate. Each chapter offers detailed information on the related subject of its research with the main objectives of the works carried out as well as providing a comprehensive list of references that should provide a rich platform of research to the field of optimum composite structures.

Chapter 1 is an introductory chapter that provides a brief review of the optimum design of composite structures as well as the relevant optimization models and techniques that are commonly implemented. As a practical application, the optimization of a composite cylindrical shell has been analyzed and solved in detail.

Chapter 2 focuses on the optimization of composite structures with nonlinear material properties. Mathematical models for flexural deformation of carbon fiber reinforced plastics and polymer matrices have been built. An application considering optimization of a multilayer composite pressure vessel is presented and discussed, where the objective function is measured by weight minimization subject to deformation and strength constraints. It is shown that the use of simplified mathematical models based on the Kirchhoff-Love and Timoshenko shell theories can be appropriate for solving the associated optimization problems.

Chapter 3 presents a model for lay-up optimization of a cantilevered composite slender, tubular beam with varied cross-section that is manufactured by winding glass fiber unidirectional tape. The multilayered composite material is assumed and modeled as a single phase anisotropic elastic homogeneous continuum. For each accepted lay-up scheme and unidirectional prepreg orientation of the symmetric balanced laminate formation, the elastic moduli were determined independently by two methods, namely; the finite element method and the classical lamination theory. The first stage is based on the analysis of the angular distribution of all engineering constants of laminates. This analysis allows the choice of a small set of "candidate" lay-ups, which are used at the modeling of the mechanical response of the beam structure at three different load scenarios. The higher level "candidates" were appointed for the final dynamic test, which includes applying full load to the selected structures and allows for the possibility to make the expert decision about final choice of quasi-optimal structure. The short discussion of the obtained results confirms the necessity of multi-objective optimization, considering many requirements and constraints that help in making the final choice of the optimal lay-up parameters.

As an important structural element in several aerospace applications, **Chapter 4** treats design sensitivity of stiffened composite panels using finite element analysis (FEA) and analytical solution models. Manufacturing and experimental measurements of a hat-stiffened composite structure is performed to validate the FEA and idealized analytical solutions. This is an attempt to initiate a structural architecture methodology to speed up the development and qualification of composite aircraft structures that will reduce design cost, increase the possibility of content reuse, and improve time-to-market. In particular, FEA results were compared with analytical solutions to develop a design methodology that will allow extensive reuse of parametric hat-stiffened panels in the design of composite structural components. This methodology is now widely utilized in developing a library of commonly used, built-in, composite structural elements in the design of modern aircrafts. The main goal of the authors is to provide the aviation industry with the most up-to-date databases for the application of advanced composite materials incorporated into parametric models to eliminate redundancies in the current process. The results include a correlated material database, an optimized model component library and a standardized way to design future complex composites structures, e.g. hat-stiffened composites panels, with reliable and predictable quality and material weight/cost.

A new hybrid material made of fiber metal laminate (*FML*), which has been successfully applied to commercial aircraft structures, is introduced in **Chapter 5**. A common type is made of Glass Reinforced Aluminum Laminate "GLARE", which combines thin aluminum sheets with unidirectional glass fiber reinforced epoxy layers. Such advanced composite material can offer weight savings of 10% compared with conventional aluminum and its alloys, together with benefits that include high tensile strength, better fatigue and damage tolerance characteristics and high level of fiber safety. A large number of practical applications demonstrate that the material properties of FMLs and their additional interlinked advantages make them the ideal option for thin-walled fuselage shells of next single aisle aircrafts. In addition, two methods have been introduced to predict the corresponding static properties with respect to the different lay-up patterns. Recently, the FML manufacturers have continued to make a substantial progress in production technology, which allows for enabling FMLs in high-volume production rates and increasing affordability for aerospace industry. In addition to the consideration of each constituent material's properties, a strong interfacial bonding between metal sheets and composite layers is one of the key factors for the improvement in joint strength and long-term durability of FML structures. Therefore, a proper surface treatment on the metallic substrate is a prerequisite for achieving long-term service capability through more efficient processing in production. Another work on *FMLs* is provided in **Chapter 6**. The main goal of the study is to assess the application of metal pins deposited by CMT (Cold Metal Transfer) PIN on metal surfaces used as layers of FMLPs, yet controlling the thickness and the number of prepreg layers. The methodological approach includes comparing small-sized FMLPs with different pin deposition patterns and spans to a reference (without pins) FMLP, in terms of energy dissipation during drop-weight testing, impact damage characterization and buckling test after impact. Iosipescu shear test, modal analysis and cosmetic characterization are also carried out.

Chapter 7 presents a method for minimizing the cost and weight of reinforced ordinary and High Strength Concrete (HSC) T-beams at limit state according to Eurocode2 (EC-2). The first objective function includes the costs of concrete, steel and formwork, whereas the second objective function represents the weight of the T-beam. All the constraint functions are set to meet the ultimate strength and serviceability requirements of Eurocode2 and current practices rules. The optimization process is developed through the use of the Generalized Reduced Gradient algorithm. Two example problems are considered in order to illustrate the applicability of the proposed design model and solution methodology. It is concluded that this approach is economically more effective compared to conventional design methods applied by structural engineers and can be extended to deal with other sections without major alterations.

Considering the next optimization of nano-composites, **Chapter 8** investigates the effect of interface on the performance of epoxy-nano clay nano-composites. The nano-clay (oMMT) filler and polymer matrix (epoxy) of the polymer nano-composites play a very important role in improving the electrical, thermal and mechanical properties. Detailed studies on the interfacial effects of filler-matrix on several properties are investigated. The chemical bonding established between epoxy and oMMT nano-filler has been investigated using Fourier Transform Infrared Spectroscopy (FTIR). The cross linking between the polymer and nano-filler was measured to determine the glassy state of the nano-composite called glass transition temperature by using Differential Scanning Calorimeter (DSC). Further, the Positron Annihilation Spectroscopy (PALS) was utilized to determine free volume as outlined in a multi-core model. In this study, a brief explanation of nano-composite interface dynamics, free volume estimation and the effect of intermolecular interactions and hydrogen bonding are included. The effect of these results on electrical property such as dielectric strength (DES) at room temperature has been thoroughly investigated.

Chapter 9 presents a variety of optimized ***FGM*** structures along with detailed structural analysis and design. The mathematical formulation is based on dimensionless quantities; therefore, the analysis is valid for different configurations and sizes. Such normalization has led to naturally scaled optimization models, which is favorable for most optimization techniques. Case studies include structural dynamic optimization of thin-walled beams in bending motion, optimization of drive shafts against torsional buckling and whirling, and aeroelastic optimization of subsonic aircraft wings. Other stability problems concerning fluid-structure interaction have also been addressed. Several design charts that are useful for direct determination of the optimal values of the design variables are introduced.

It is hoped that this book will prove of particular value to structural engineers and researchers working in the field. It should also prove useful to postgraduates wishing to gain special knowledge on design optimization of composite structures. Finally, I am glad to have had the opportunity of acknowledging all the contributing authors and express my gratitude for the help and support of INTECH staff particularly the Author Service Manager **Ms. Marijana Francetic**.

Karam Y. Maalawi
Professor of Aeronautics & Mechanics
National Research Centre
Cairo, Egypt

Chapter 1

Introductory Chapter: An Introduction to the Optimization of Composite Structures

Karam Maalawi

Additional information is available at the end of the chapter

http://dx.doi.org/10.5772/intechopen.81165

1. Introduction

Structural applications of composite materials are increasing in several engineering areas where high stiffness and strength-to-weight ratios, long fatigue life, superior thermal properties, and corrosive resistance are most beneficial [1–4]. Common types include laminated composites [5], functionally graded material (*FGM*) structures, and nanocomposites as well as smart composite structures [6]. In fact composite structures are usually tailored, depending upon the specific objectives, by choosing the individual constituent materials and their volume fractions, fiber orientation angles, and laminas thickness and number, as well as the fabrication procedure. To attain the best results, adequate optimization models have to be implemented to find practical optimal solutions satisfying a given set of design constraints.

This introductory chapter provides a brief review on the optimum design of composite structures and the relevant optimization techniques that are capable of finding the needed optimal solutions. Several problems can be addressed, including the structural design for maximum stability, maximum natural frequencies, and minimum mass or maximum stiffness subject to limits on strength, deflections, and side constraints. The relevant design variables include geometrical dimensions and material properties as well. A numerical example is given at the end of this chapter to demonstrate a real and practical application of the optimum composite structures.

2. The optimal design problem

Several research papers and text books exist in the field of optimal design of composite structures with a variety of valuable applications in civil, mechanical, ocean, and aerospace engineering. An important stage has now been reached at which an investigation of such developments and their practical possibilities should be made and presented. Two distinct review papers have been published covering the development of the optimum design of composites over more than 40 years. The first paper by Sonmez [7] presented a comprehensive survey for more than 1000 journal papers, conference papers, textbooks, and web links from the year 1969 to 2009. Sonmez classified the papers according to the type of the composite structure, loading conditions, optimization model, failure criteria, and the utilized search algorithm. The second paper by Ganguli [8] covered a historical review from 1973 to 2013. It provides the growth of the field by including more than 90 references dealing with a variety of optimization methods utilized for tailoring composites to achieve certain design objectives. Applications of several optimization techniques were presented, including feasible direction methods, sequential quadratic programming, and stochastic optimization such as particle swarm and ant colony algorithms. Ganguli classified the published work into five categories named pioneering research for the work published in the 1970s, early research in the 1980s, moving toward design in the 1990s, the new century in the 2010s, and the current research for papers published after 2010.

In general, design optimization seeks the best values of **design variables,** $\underline{\mathbf{X}}_{\mathbf{nx1}}$, to achieve, within certain **constraints,** $\underline{\mathbf{G}}_{\mathbf{mx1}}(\underline{\mathbf{X}})$ placed on the system behavior, allowable stresses, geometry, or other factors; its goal of optimality is defined by the a vector of **objective functions,** $\underline{\mathbf{F}}_{\mathrm{kx1}}(\underline{\mathbf{X}})$, for specified environmental conditions. Mathematically, design optimization may be cast in the following standard form [9]:

Find the set of design variables $\underline{X}_{nx1}$ that will

$$\text{minimize} \quad F(\underline{X}) = \sum_{i=1}^{k} w_{fi} F_i(\underline{X}) \tag{1}$$

$$\text{subject to} \quad G_j(\underline{X}) \leq 0, j = 1, 2, \ldots \mathrm{I} \tag{2}$$

$$G_j(\underline{X}) = 0, j = I + 1, I + 2, \ldots m \tag{3}$$

where w_{fi} is the weighting factors measuring the relative importance of $F_i(x)$ with respect to the overall design goal:

$$0 \leq w_{fi} \leq 1 \qquad \sum_{i=1}^{k} w_{fi} = 1 \tag{4}$$

Figure 1 shows the overall structure of an optimization approach to design. Major objectives in mechanical and structural engineering involve minimum fabrication cost, maximum product reliability, maximum stiffness/weight ratio, minimum aerodynamic drag, maximum natural frequencies, maximum critical shaft speeds, etc. Design variables describe configuration, dimensions and sizes of elements, and material properties as well. In the design of structural components, such as those of an automobile structure, the main design variables represent the thickness of the covering skin panels and the spacing, size, and shape of the transverse and

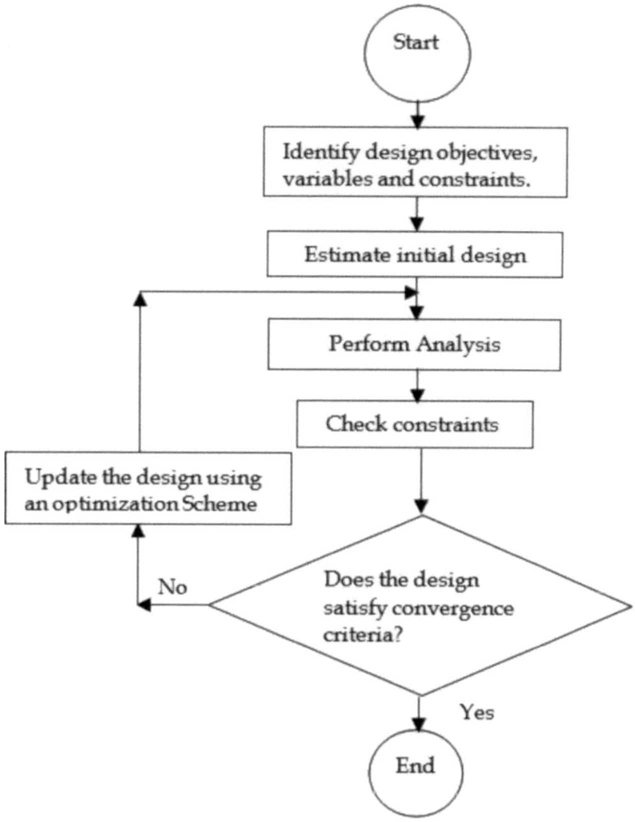

Figure 1. Design optimization process.

longitudinal stiffeners. The sizes of the constituent elements of the system are measured by such properties as the cross-sectional dimensions, section areas, area moments of inertia, torsional constants, plate's thicknesses, etc. If the skin and/or stiffeners are made of layered composites, the orientation of the fibers and their proportion can become additional variables. If one optimizes for configuration, the design variables will include spatial coordinates. Also, in dynamic problems, the location of nonstructural masses and their magnitudes can be additional design variables.

3. Optimization techniques

The class of optimization problems described by Eqs. (1)–(3) may be thought of as a search in an n-dimensional space for a point corresponding to the minimum value of the overall objective function and such that it lies within the region bounded by the subspaces representing the constraint functions. Iterative techniques are usually used for solving such optimization problems in which a series of directed design changes (moves) are made between successive points in the design space. Several optimization techniques are classified according to the way of selecting the search direction [9]. The most commonly used approaches are the random search, conjugate directions, and conjugate gradients methods. Other algorithms for solving global optimization problems may be classified into heuristic methods that find the global optimum only with high probability and methods that guarantee to find a global optimum with some accuracy. The simulated annealing technique and the genetic algorithms (GAs) belong to the former type, where analogies to physics and biology to approach the global optimum are utilized. The simulated annealing technique is an iterative search method based on the simulation of thermal annealing of critically heated solids. Hasancebi et al. [10] applied it to find the optimum design of fiber composite structures as an efficient method to solve multi-objective optimization models. On the other hand, the GAs [11, 12] are based on the principles of natural genetics and natural selection. GAs do not utilize any gradient information during the searching process. Narayana Naik et al. [12] used GA and various failure mechanisms based on different failure criteria to reach an optimal composite structure. Another robust algorithm in solving complex problems of optimal structural design is named particle swarm optimization algorithm (PSOA). This algorithm is based on the behavior of a colony of living things, such as a swarm of insects like ants, bees, and wasps, a folk of birds, or a school of fish. Omkar et al. [13] applied PSOA to achieve a specified strength with minimizing weight and total cost of a composite structure under different failure criteria. To the author's knowledge, GA has been the most efficient stochastic method for obtaining the global optimum design of composite structures.

4. Application: buckling optimization of anisotropic cylindrical shells

Structural buckling failure due to high external hydrostatic pressure is a major consideration in designing cylindrical shell-type structures. This section presents a direct approach for enhancing buckling stability limits of thin-walled long cylinders that are fabricated from multi-angle fibrous laminated composite lay-ups. The mathematical formulation employs the classical lamination theory for calculating the critical buckling pressure, where an analytical solution that accounts for the effective axial and flexural stiffness separately as well as the inclusion of the coupling stiffness terms is presented. The associated design optimization problem of maximizing the critical buckling pressure has been formulated in a standard nonlinear mathematical programming problem with the design variables encompassing the fiber orientation angles and the ply thicknesses as well. The physical and mechanical properties of the composite material are taken as preassigned parameters. The proposed model deals with dimensionless quantities in order to be valid for thin shells having different thickness-to-radius ratios. Results have been obtained for cases of filament wound cylinders fabricated from different types of composite materials.

The basic analysis and analytical formulation presented in this chapter are based on the work given by Maalawi [14], which provides good sensitivity to lamination parameters and allows the search for the needed optimal stacking sequences in a reasonable computational time. Referring to the structural model depicted in **Figure 2**, the significant strain components are

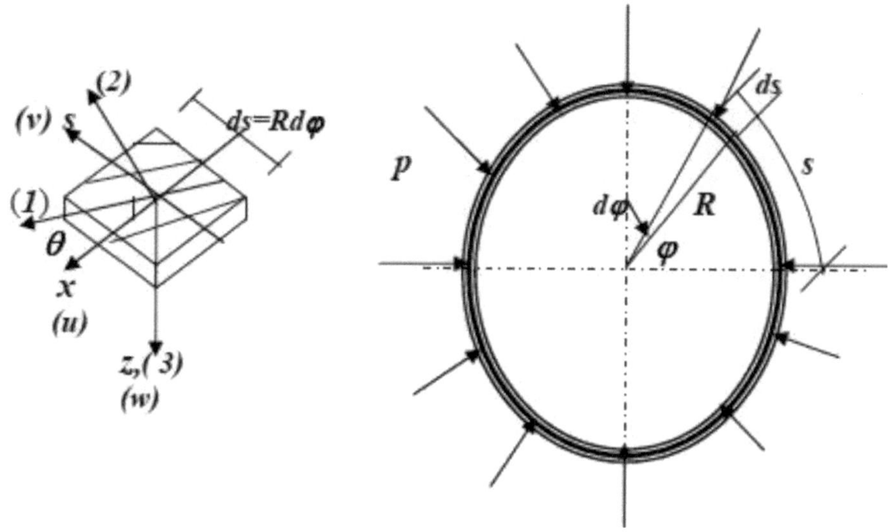

Figure 2. Laminated composite cylindrical shell under external pressure (*u* displacement in the axial direction *x*, *v* in the tangential direction *s*, *w* in the radial direction *z*).

the hoop strain (ε_{ss}^{0}) and the circumferential curvature (*Kss*) of the mid-surface. The reduced form of the stress-strain relationships in matrix form is

$$\begin{Bmatrix} N_{ss} \\ M_{ss} \end{Bmatrix} = \begin{bmatrix} A_{22} & B_{22} \\ B_{22} & D_{22} \end{bmatrix} \begin{Bmatrix} \varepsilon_{ss}^{0} \\ \kappa_{ss} \end{Bmatrix} \tag{5}$$

where N_{ss} and M_{ss} are the resultant distributed force and moment and (A_{ij}, B_{ij}, D_{ij}) are the extensional, coupling, and bending stiffness coefficients, respectively [1].

4.1. Analytical buckling model

The governing differential equations of anisotropic long cylinders subjected to external pressure are cast in the following:

$$M'_{ss} + R\left(N'_{ss} - \beta N_{ss}\right) = \beta\, pR^2 \tag{6.1}$$

$$M''_{ss} - R\left[N_{ss} + \left(\beta N_{ss}\right)' + p\left(w_o + v'_o\right)\right] = pR^2 \tag{6.2}$$

where u_o, v_o, *and* w_o are the displacements of a generic point (x, s) on the shell middle surface ($z = 0$) in x, s, and z directions, respectively. The prime denotes differentiation with respect to the angular position φ and $\beta = \left(v_o - w'_o\right)/R$. For the case of thin cylinders with thickness-to-radius ratio (h/R) ≤ 0.1, the critical buckling pressure can be determined using the mathematical expression [14]:

$$p_{cr} = 3\left[\frac{D_{22}}{R^3}\right]\left[\frac{1-(\psi^2/\alpha)}{1+\alpha+2\psi}\right] \quad (7.1)$$

$$\psi = \left(\frac{1}{R}\right)\left(\frac{B_{22}}{A_{22}}\right) \quad (7.2)$$

$$\alpha = \left(\frac{1}{R^2}\right)\left(\frac{D_{22}}{A_{22}}\right) \quad (7.3)$$

4.2. Definition of the baseline design

It is convenient first to normalize all variables and parameters with respect to a baseline design, which has been selected to be a unidirectional orthotropic laminated cylinder with the fibers parallel to the shell axis x. Optimized designs shall have the same material properties, mean radius R, and total shell thickness h of the baseline design. Expressions for calculating the critical buckling pressure (P_{cro}) of the baseline design are defined in **Table 1**, which depend upon the type of composite material utilized and the shell thickness-to-radius ratio (h/R) as well.

4.3. Optimization model

The search for the optimized lamination can be performed by coupling the analytical buckling shell model to a standard nonlinear mathematical programming procedure. The resulting optimization problem may be cast in the following standard form to

$$\text{minimize} \quad -\hat{p}_{cr} \quad (8.1)$$

$$\text{subject to} \quad h_L \leq \hat{h}_k \leq h_U, \quad (8.2)$$

$$\theta_L \leq \theta_k \leq \theta_U \qquad k = 1, 2, \ldots.n \quad (8.3)$$

$$\sum_{k=1}^{n} \hat{h}_k = 1 \quad (8.4)$$

where $\hat{p}_{cr} = p_{cr}/p_{cro}$ is the dimensionless critical buckling pressure, and (h_L, h_U) are the lower and upper bounds imposed on the individual dimensionless ply thicknesses $\hat{h}_k = h_k/h$.

Material type	Orthotropic mechanical properties* (GPa)				$P_{cro} \times (h/R)^3$ (GPa)
	E_{11}	E_{22}	G_{12}	ν_{12}	
E-Glass/vinyl ester	41.06	6.73	2.5	0.299	1.708
Graphite/epoxy	130.0	7.0	6.0	0.28	1.757
S-Glass/epoxy	57.0	14.0	5.7	0.277	3.567

*E_{11} = longitudinal modulus, E_{22} = hoop modulus, ν_{12} = Poisson's ratio for axial load, $\nu_{21} = \nu_{12}E_{22}/E_{11}$.

Table 1. Material properties and critical buckling pressure of the baseline design (P_{cro}).

According to the filament-winding manufacturing process, each ply is characterized by its angle θ_k with respect to the cylinder axis x. The stacking sequence is denoted by $[\theta_1/\theta_2/\ldots/\theta_n]$, where the angles are given in degrees, starting from the outer surface of the shell. In addition, in a real-world manufacturing process, the filament-winding angles θ_k must be chosen from a limited range of allowable lower (θ_L) and upper (θ_U) values according to technology references. It is important to mention here that the volume fractions of the constituent materials of the composite structure is assumed to not significantly change during optimization, so that the total structural mass remains constant at its reference value of the baseline design.

4.4. Optimal solutions

The functional behavior of the candidate objective function, as represented by maximization of the dimensionless buckling pressure $\hat{p}_{cr}$, is thoroughly investigated in order to see how it is changed with the optimization variables in the selected design space. The final optimum designs recommended by the model will directly depend on the mathematical form and behavior of the objective function.

4.4.1. Two-layer anisotropic long cylinder

The first case study to be considered herein is a long thin-walled cylindrical shell fabricated from E-glass/vinyl ester composites with the lay-up composed of only two plies ($n = 2$) having equal thicknesses ($\hat{h}_1 = \hat{h}_2 = 0.5$) and different fiber orientation angles. Considering the case of $\pm 63°$ angle ply, the present model gives $\hat{p}_{cr} = 4.23$, *i.e.*, $P_{cr} = 4.23 \times 1.708 \times (h/R)^3$ *GPa*, depending on the shell thickness-to-radius ratio. The actual dimensional values of the critical buckling pressure for the different thickness ratios are given in **Table 2** for the cases of baseline design [0°], helically wound [±63°], and [±90°] hoop layers. The unconstrained maximum value of $\hat{p}_{cr} = 6.1$ occurs at the design points $[\theta_1/\theta_2] = [\pm 90, \pm 90]$.

For a two-ply long cylinder fabricated from graphite/epoxy composites, **Figure 3** shows the developed level curves of the dimensionless buckling pressure, $\hat{p}_{cr}$ (also named isomerits or isobars) in the $(\theta_1 - \theta_2)$ design space. As seen in the figure, the maximum value of $\hat{p}_{cr}$ reaches a value of *18.57* for a hoop wound construction. **Table 3** presents the solutions for the [±45°] angle-ply and the [90°] cross-ply constructions for different thickness-to-radius ratios. These solutions

	Baseline [0°]	**Helically wound [±63°]**	**Hoop plies [±90°]**
	$\hat{p}_{cr} = 1.00$	**4.23**	**6.10**
(h/R)			
(1/15)	506.07	2140.69	3087.05
(1/20)	213.50	903.11	1302.35
(1/25)	109.31	462.39	666.80
(1/50)	13.66	57.80	83.35

$[P_{cr} = \hat{p}_{cr} \times 1.708 \times 10^6\ (h/R)^3$ KPa].

Table 2. Critical buckling pressure for E-glass/vinyl ester cylinder with different lay-ups.

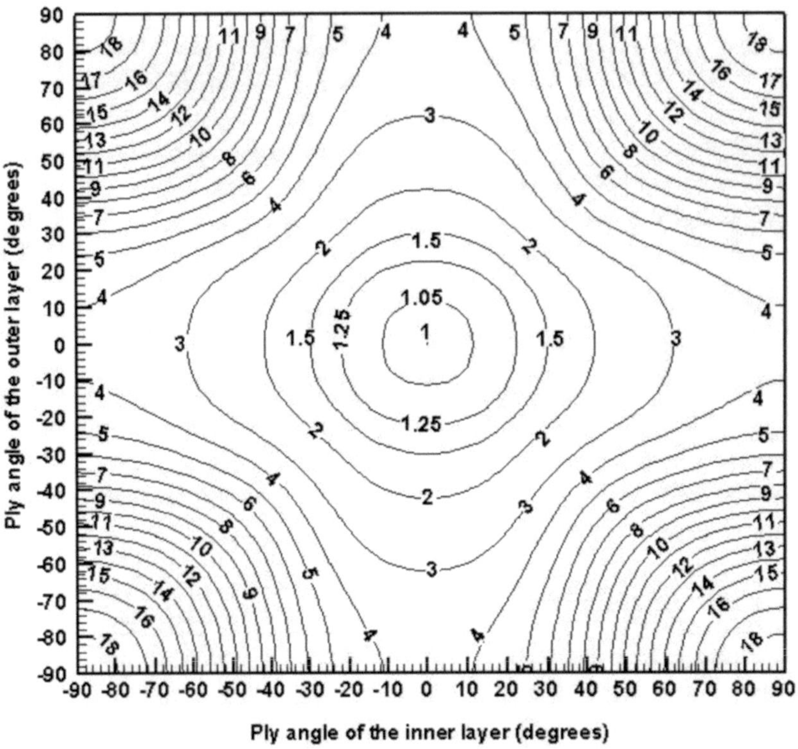

Figure 3. $\hat{p}_{cr}$-isomerits for a graphite/epoxy, two-layer cylinder in $[\theta_1/\theta_2]$ design space ($\hat{h}_1 = \hat{h}_2 = 0.5$).

are also valid for lay-ups $[0_3°]_s$, $[90_3°]_s$, $[45_2°/-45_2°]_s$, and $[45°/-45°/45°/-45°]_s$. The case of a helically wound lay-up construction $[+\theta/-\theta]$ with unequal play thicknesses $\hat{h}_1$ and $\hat{h}_2$, such that their sum is held fixed at a value of unity, has also been investigated. Computer solutions have shown that no significant change in the resulting values of the critical buckling pressure can be remarked in spite of the wide change in the ply thicknesses. This is a natural expected result since the stiffness coefficients A_{22}, B_{22}, and D_{22} remain unchanged for such lay-up construction.

	Baseline [0°]	**Helically wound [±45°]**	**Hoop plies [±90°]**
	$\hat{p}_{cr}$ **= 1.00**	**5.9**	***18.57***
(h/R)			
(1/15)	520.59	3071.50	9667.40
(1/50)	14.06	82.93	261.02
(1/120)	1.02	5.99	18.88

$[P_{cr} = \hat{p}_{cr} \times 1.757 \times 10^6 (h/R)^3$ KPa].

Table 3. Critical buckling pressure, P_{cr} for graphite/epoxy cylinder with different lay-ups.

	Baseline [$0°_3$]	[0°/90°/0°]	[90°/0°/90°]
	$\hat{p}_{cr}$ = **1.00**	**1.651**	**17.92**
(h/R)			
(1/15)	520.59	859.57	9331.19
(1/50)	14.06	23.21	251.94
(1/120)	1.02	1.68	18.23
[$Pcr = \hat{p}_{cr} \times 1.757 \times 10^6 (h/R)^3$ KPa].			

Table 4. Critical buckling pressure, P_{cr} for graphite/epoxy cylinder [$\theta_1/\theta_2/\theta_1$].

4.4.2. *Three-layer anisotropic long cylinder*

Results for a cylinder constructed from three, equal-thickness layers with stacking sequence denoted by [$\theta_1/\theta_2/\theta_1$] are given in **Table 4**. The same behavior can be observed as before but with slight change in the attained values. It was found that for the range $-30° > \theta_1 > 30°$ the critical buckling pressure is not much affected by variation in the ply angle θ_2. A substantial increase in the critical buckling pressure by changing the ply angles can be observed. Similar solutions were obtained for the stacking sequences $[0°_2/90°]_s$ and $[90°_2/0°]_s$.

4.4.3. *Four-layer sandwiched anisotropic cylinder*

The same graphite/epoxy cylinder is reconsidered here with changing the stacking sequence to become ±20° equal-thickness layers sandwiched in between outer and inner 90° hoop layers with unequal thicknesses, i.e., ($\hat{h}_2 = \hat{h}_3$) and ($\hat{h}_1 \neq \hat{h}_4$), such that the thickness equality constraint $\sum_{k=1}^{4} \hat{h}_k=1$ is always satisfied. **Figure 4** shows the developed $\hat{p}_{cr}$ -isomerits in the ($\hat{h}_1, \hat{h}_2$) design space. The contours inside the feasible domain, which is bounded by the three lines $\hat{h}_1 = 0$ and $\hat{h}_2 = 0$ and $\hat{h}_1 + 2\hat{h}_2 = 1$ (i.e., $\hat{h}_4 = 0$), are obliged to turn sharply to be asymptotes to the line $\hat{h}_4 = 0$, in order not to violate the thickness equality constraint. This is why they appear in the figure as zigzagged lines. At the design point ($\hat{h}_1, \hat{h}_2$) = (0.25, 0.25), the dimensionless buckling pressure $\hat{p}_{cr}$ = 16.43 (see **Figure 4** and **Table 5**). As a general observation, as the thickness of the hoop layers increases, a substantial increase in the critical buckling pressure will be achieved, e.g., at ($\hat{h}_1, \hat{h}_2$) = (0.33, 0.17), $\hat{p}_{cr}$= 17.92 representing a percentage increase of $(17.92 - 16.43)/16.43 = 9.1\%$.

Finally, the obtained results have indicated that the optimized laminations induce significant increases, always exceeding several tens of percent, of the buckling pressures with respect to the reference or baseline design. It is assumed that the volume fractions of the composite material constituents do not significantly change during optimization, so that the total structural mass remains constant. It has been shown that the overall stability level of the laminated composite shell structures under considerations can be substantially improved by finding the optimal stacking sequence without violating any imposed side constraints. The stability limits

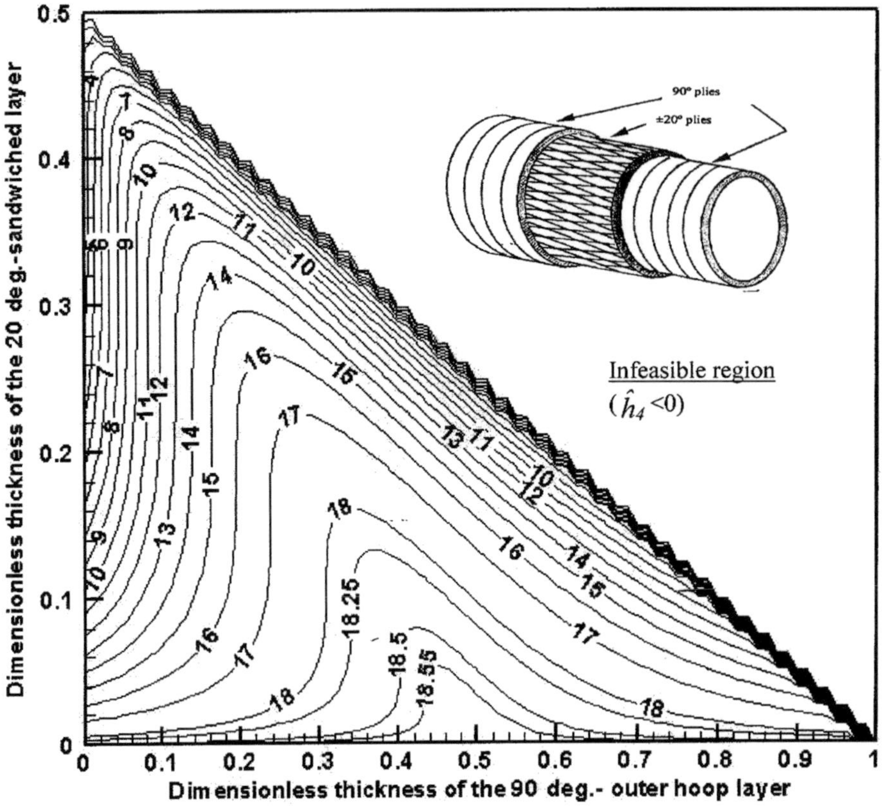

Figure 4. Design space for a sandwich lay-up graphite/epoxy cylinder [*90/±20/90*].

(*h/R*)	(*1/15*)	(*1/20*)	(*1/25*)	(*1/50*)
Pcr	8553.0	3609.3	1847.5	231.0
[$\hat{p}_{cr} = 16.43$, $P_{cro} = 1.757 \times 10^6\ (h/R)^3$ KPa, $Pcr = \hat{p}_{cr} \times P_{cro}$].				

Table 5. Critical buckling pressure, P_{cr} (KPa), for graphite/epoxy cylinder [*90/±20/90*].

of the optimized shells have been substantially enhanced as compared with those of the reference or baseline designs.

Author details

Karam Maalawi

Address all correspondence to: maalawi@netscape.net

National Research Centre, Cairo, Egypt

References

[1] Daniel I, Ishai O. Engineering Mechanics of Composite Materials. 2nd ed. Oxford University Press, New York; 2006. 432p. ISBN: 978-0195150971

[2] Kollar LP, Springer GS. Mechanics of Composite Structures. 1st ed. United Kingdom: Cambridge University Press; 2009. 500p. ISBN: 978- 0521126908

[3] Niu MCY. Composite Airframe Structures. 3rd ed. Hong Kong: Hong Kong Conmilit Press Ltd; 2011. 500p. ISBN: 978-9627128069

[4] Kassapoglou C. Design and Analysis of Composite Structures: With Applications to Aerospace Structures. 2nd ed. New Jersey: Wiley; 2013. 410p. ISBN: 978-1118401606

[5] Ye J. Laminated Composite Plates and Shells, 3D Modelling. 1st ed. London: Springer-Verlag; 2003. DOI: 10.1007/978-1447100959. 273p. ISBN: 978-1447100959

[6] Kumar SS. Smart Composite Structures. 1st ed. United States: Lambert Academic Publishing; 2013. 180p. ISBN: 978-3659322532

[7] Sonmez FO. Optimum design of composite structures: A literature survey (1969–2009). Journal of Reinforced Plastics and Composites. 2017;**36**(1):3-39. DOI: 10.1177/0731684416668262

[8] Ganguli R. Optimal design of composite structures: A historical review. Journal of the Indian Institute of Science. 2013;**93**(4):557-570. ISSN: 0970-4140 Coden-JIISAD

[9] Maalawi K, Badr M. Design optimization of mechanical elements and structures: A review with application. Journal of Applied Sciences Research. 2009;**5**(2):221-231

[10] Hasancebi O, Carba S, Saka M. Improving the performance of simulated annealing in structural optimization. Structural and Multidisciplinary Optimization. 2010;**41**:189-203

[11] Walker M, Smith RE. A technique for the multi-objective optimization of laminated composite structures using genetic algorithms and finite element analysis. Composite Structures. 2003;**62**:123-128

[12] Narayana Naik G, Gopalakrishnan S, Ganguli R. Design optimization of composites using genetic algorithms and failure mechanism based failure criterion. Composite Structures. 2008;**83**:354-367

[13] Omkar SN, Senthilnath J, Khandelwal R, Narayana Naik G, Gopalakrishnan S. Artificial bee colony (ABC) for multi-objective design optimization of composite structures. Applied Soft Computing. 2011;**11**(1):489-499

[14] Maalawi KY. Optimal buckling design of anisotropic rings/long cylinders under external pressure. Journal of Mechanics of Materials and Structures. 2008;**3**(4):775-793

Chapter 2

Mathematical Modeling and Numerical Optimization of Composite Structures

Sergey Golushko

Additional information is available at the end of the chapter

http://dx.doi.org/10.5772/intechopen.78259

Abstract

This chapter is devoted to modeling the properties of composite materials and structures. Mathematical relations describing the nonlinear elastic three-point bending of isotropic and reinforced beams with account of different strength and stiffness behavior in tension and compression are obtained. An algorithm for numerical solution of corresponding boundary-value problems is proposed and implemented. Results of numerical modeling were compared to acquired data for polymer matrix and structural carbon fiber reinforced plastics. A computational technology for analysis and optimization of composite pressure vessels was developed and presented.

Keywords: composite, polymer matrix, CFRP, bending, nonlinear deformation, mathematical modeling, pressure vessel, COPV, shell theory, optimization

1. Introduction

Carbon fiber reinforced plastics (CFRP) are the most promising modern composite materials. High-duty structures used in aviation and space industry, car manufacturing and building sector require new CFRPs as well as ways to improve their characteristics. Applying computer modeling techniques significantly reduces both the time and cost of investigations aimed at searching optimal parameters of CFRP structures [1]. Mathematical modeling provides an opportunity for comprehensive analysis of both CFRPs and CFRP structures. It has become an effective tool for solving important applied problems.

To build a mathematical model of composite materials, including those made of carbon fibers, one relies upon the experimental data acquired in mechanical testing. Wide application of digital testing machines has brought such experiments to higher level of quality. By measuring a large number of parameters with a high discretization frequency, modern testing machines allow for high amount of information on material deformation and failure to be obtained within a single experiment. Therefore, data processing has become an important step for mathematical modeling of CFRPs and CFRP structures. This is exceptionally important because of quite specific behavior of CFRPs and of their components: fibers and matrices.

One of the features of such materials is their different strength and stiffness behavior in tension and compression combined with nonlinearities of stress-strain curves. Multiple studies for epoxy matrices showed that their ultimate strains in tension were much lower than in compression: approximately 4 versus 20% and more [2]. Moreover, under tension and compression, the deformation behavior of epoxy matrices significantly differs. The corresponding stress-strain curves have different stiffness (secant modulus) at the same values of strain. The similar difference can be observed for CFRPs. In [3–5], it was shown that in tension tests of carbon fiber specimens with reinforcement angles less than 20° the stiffness grows together with strains (stiffening), whereas for epoxy matrices softening is observed. This phenomenon was explained by the properties of carbon fibers.

Contrasting behavior in tension and compression, stiffening, softening and other nonlinearities are forcing researchers to build and use special mathematical models and computing algorithms. Mathematical models taking into account the abovementioned properties of materials were proposed and studied theoretically by Timoshenko [6] and Ambartsumyan [7, 8] in the mid-twentieth century. Later, Jones had experimentally, theoretically and numerically studied the nonlinear behavior of several fiber-reinforced composites. The main focus of the research was on the difference in stiffness and strength behavior under tension and compression [9]. After Ambartsumyan's and Jones' researches, a lot of studies were dedicated to this problem. Most of them were dealing with linear bi-modulus models of materials or 3D finite elements. In [10], Ambartsumyan with a coauthor suggested a theoretical approach to modeling of multimodulus nonlinear elastic beams under bending, but still without calculations.

Another trend is studying the behavior of sandwich panels or beams with a CFRP faces having differences in tension and compression along with the other mentioned nonlinearities [11–13]. These works concern the problem of flexure of CFRPs and similar materials. They consider bending of specimens as a reference test. The first paper [11] is devoted to experimental investigations and shows most of the nonlinearities we supposed such materials should have: stiffening in tension, softening in compression, different moduli even at the origin of coordinates. Other two works [12, 13] present more complex studies including full cycle of mathematical modeling spanning from the experimental investigations to numerical ones.

A comprehensive approach to modeling and simulation of nonlinear elastic deformation of polymer matrices and different CFRPs was presented in [14]. This chapter deals with different strength and stiffness behavior of the materials in tension and compression exemplified by a case of three-point bending. This approach implements a full cycle of model development and validation, which comprises the following stages: carrying out tests and acquiring experimental data, data prepossessing and building stress-strain curves, analytical approximation of acquired curves, mathematical modeling and numerical simulation of deformation processes, comparative analysis of results of numerical modeling to acquired data.

2. Structural models of composite materials

For most of the composite materials models, we can write the relations between average stresses $\sigma_{\alpha\beta}$, $\tau_{\alpha 3}$ and strains $e_{\alpha\beta}$, $\gamma_{\alpha 3}$ (generalized Hooke's Law):

$$\begin{aligned} \sigma_{\alpha\alpha} &= a_{\alpha\alpha} e_{\alpha\alpha} + a_{\alpha\beta} e_{\beta\beta} + a_{\alpha 3} \cdot 2e_{\alpha\beta} - a_{\alpha\Theta}\Theta, \\ \sigma_{\alpha\beta} &= a_{\alpha 3} e_{\alpha\alpha} + a_{\beta 3} e_{\beta\beta} + a_{33} \cdot 2e_{\alpha\beta} - a_{3\Theta}\Theta, \\ \gamma_{\alpha 3} &= q_{\alpha\alpha} \tau_{\alpha 3} + q_{\alpha\beta} \tau_{\beta 3}, \end{aligned} \tag{1}$$

where Θ is the increase of temperature. Relations (Eq. (1)) are called the thermoelasticity relations, or, when no temperature influence is considered, they are simply elasticity relations.

The structural model of fiber reinforced composite described in [15–18] has become a foundation for a large number of current researches. Now it is widely used while simulating the behavior of composite structures. The model is based on the following assumptions: the stress-strain state into isotropic elastic fibers and into entire volume of isotropic ideally elastic matrix is homogeneous; fibers and matrix are deformed jointly along the direction of reinforcement; stresses in fibers and in matrix corresponding to other directions are equal.

For computing the effective elastic modulus of unidirectional fiber-reinforced composite, the Reuss-Voigt average was used giving the following formulae

$$\begin{aligned} E_1 &= \omega_f E_f + \omega_m E_m, \quad E_2 = \frac{E_f E_m}{\omega_f E_m + \omega_m E_f}, \\ \nu_{12} &= \omega_f \nu_f + \omega_m \nu_m, \quad G = \frac{G_f G_m}{\omega_f G_m + \omega_m G_f}, \end{aligned} \tag{2}$$

where all the terms having squared Poisson coefficients are neglected.

Herewith E_1, E_2 are effective moduli along and across the direction of reinforcement, G is effective share modulus, ν_{12} is effective Poisson coefficient in the plain of layer; E, ν, ω with "f" and "m" indices are elastic moduli, Poisson coefficients and volume fractions of matrix and fibers correspondingly, hereby $\omega_m + \omega_f = 1$.

On the ground of symmetry of compliance tensor, one has

$$\nu_{21} = \nu_{12} E_2 E_1^{-1}.$$

Formulae for effective coefficients of thermal expansion have the following form

$$\alpha_1 = \omega_f \alpha_f + \omega_m \alpha_m, \quad \alpha_2 = \frac{\omega_f \alpha_f E_f + \omega_m \alpha_m E_m}{\omega_f E_f + \omega_m E_m}. \tag{3}$$

In description of the model, it is noted that among formulae for effective moduli, those obtained using Reuss averaging (in particular formulae for G) lead to the worst results.

Estimations for G obtained using variational method are also obtained, and it is shown that lower boundary

$$G = \frac{(1+\omega_f)G_f + \omega_m G_m}{(1+\omega_f)G_m + \omega_m G_f} G_m \tag{4}$$

gives more accurate approximation than (Eq. (2)) does. Hereinafter share moduli of matrix and fibers are

$$G_m = \frac{E_m}{2(1+\nu_m)}, \quad G_f = \frac{E_f}{2(1+\nu_f)}.$$

Components of effective stiffness tensor for unidirectionally reinforced layer in case of state of plane stress have the following form:

$$A_{\alpha\alpha\alpha\alpha} = \frac{E_\alpha}{1-\nu_{12}\nu_{21}}, \quad A_{1122} = \frac{\nu_{21}E_1}{1-\nu_{12}\nu_{21}}, \quad A_{1212} = G. \tag{5}$$

Unwritten expressions can be obtained using symmetry rule or vanish. Hereinafter, we assume $\alpha, \beta = 1, 2$ and $\alpha \neq \beta$.

The coefficients in the relations (Eq. (1)) for example are defined by the formulas given in [1, 17, 18].

3. Mathematical model and numerical analysis of reinforced beam deformation

Three-point bending flexural test has been one of the standard techniques to determine physical and mechanical characteristics of materials. **Figure 1** shows a scheme of physical model of three-point bending of a beam with the rectangular cross section $b \times 2h$, and the span l between the supports. The left edge of the beam is hinged, while the right one is supported freely. The force P is applied to the center of the beam. The model neglects the shape of the supports and assumes the occurring load P and support reactions R_A and R_B to be concentrated. In addition, the model neglects the possible heterogeneity of the deformations in the direction normal both to the longitudinal direction and to the load direction.

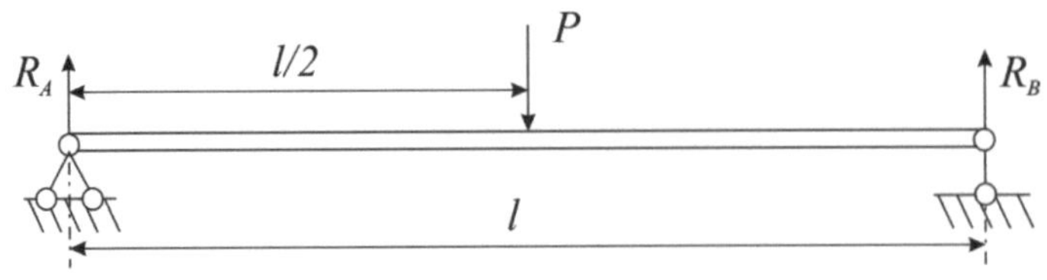

Figure 1. Three-point bending of a rectangular-sectioned beam.

In this case, the beam's upper part undergoes compression strain in the longitudinal direction, bottom part—tension strain. VSE-1212 polymer matrix and VKU-28 (T-800 carbon yarn plus VSE-1212 epoxy matrix) structured CFRP react differently to tension and compression. VKU-28 has been one of the most promising types of CFRPs that is going to be used in the latest generations of aircrafts. The effect of accounting for this factor on the computational results is essential. Further, these results are compared to acquired ones.

Due to very low deformation rates, the classical theory of beam bending can be regarded as satisfactory for description of the equilibrium state. To this end, it is convenient to consider the beam's median surface as a reference one.

The beam's stress-strain state is characterized by the following values determined on the reference surface: the shear force $Q(x)$, the bending moment $M(x)$, the longitudinal force $N(x)$, and by the longitudinal displacement and bend ($u(x)$, $w(x)$ respectively). The corresponding equilibrium equations are written as follows:

$$\frac{dN}{dx} = 0, \qquad \frac{dQ}{dx} = 0, \qquad \frac{dM}{dx} = Q. \tag{6}$$

The reactions R_A and R_B can be determined by considering force equilibrium $R_A = R_B = P/2$. The bending moments at the support points are equal to zero: $M_A = M_B = 0$. The solution of the equation system (Eq. (6)) can be expressed as follows:

$$N = 0, \quad Q(x) = \begin{cases} P/2, & 0 \leq x \leq l/2, \\ -P/2, & l/2 \leq x \leq l, \end{cases} \qquad M(x) = \begin{cases} Px/2, & 0 \leq x \leq l/2, \\ -P(x-l)/2, & l/2 \leq x \leq l. \end{cases} \tag{7}$$

Strain distribution for the beam's thickness can be obtained from the Kirchhoff-Love kinematic hypotheses:

$$\varepsilon(x,z) = e(x) + z\kappa(x), \tag{8}$$

$$e(x) = \frac{du}{dx}, \qquad \kappa(x) = -\frac{d^2w}{dx^2}, \tag{9}$$

where $\varepsilon(x,z)$ is the strain in the beam; $e(x)$ is the median surface strain; and $\kappa(x)$ denotes changes in the median surface curvature. As mentioned earlier, the beam undergoes tension and compression strain, whose interface will be marked as z_1. In this case, for the section area $-h \leq z \leq z_1$, the strain will be negative, and for $z_1 \leq z \leq h$ positive. At the interface of these two states, the strains ε vanish, so the interface itself is determined as follows:

$$z_1 = -\frac{e}{\kappa}, \qquad -h \leq z_1 \leq h. \tag{10}$$

The constitutive equation can be expressed as:

$$\sigma^{\pm}(x,z) = f_i^{\pm}(\varepsilon), \tag{11}$$

where the superscript "+" refers to the areas with positive strains and "–" – to the area with negative ones; $f(\varepsilon)$ denotes the approximation selected for the stress-strain curve (a linear function, a polynomial, or a combination of linear and power-law functions).

The longitudinal force N and the bending moment M in the beam cross section are determined by the equations:

$$\begin{aligned} N &= b\left(\int_{-h}^{z_1} \sigma^- dz + \int_{z_1}^{h} \sigma^+ dz\right), \\ M &= b\left(\int_{-h}^{z_1} \sigma^- z dz + \int_{z_1}^{h} \sigma^+ z dz\right). \end{aligned} \tag{12}$$

Having substituted (Eq. (12)) with the relations (Eq. (8)), (Eq. (10)), (Eq. (11)) into and integrated it over the beam thickness, one obtains a system of equations to determine κ and e:

at $0 \le x \le l/2$

$$\begin{cases} N(\kappa, e, x) = 0, \\ M(\kappa, e, x) = Px/2, \end{cases} \tag{13}$$

at $l/2 < x \le l$

$$\begin{cases} N(\kappa, e, x) = 0, \\ M(\kappa, e, x) = -P(x - l)/2. \end{cases} \tag{14}$$

The system of equations (Eq. (13)) and (Eq. (14)) in general case is nonlinear, but in the case of piecewise linear constitutive equations which take into account different strength and stiffness behavior in tension and compression expressed as follows:

$$\sigma^{\pm}(x, z) = E^{\pm}\varepsilon, \tag{15}$$

it can be solved analytically. In the nonlinear case, the Newton method is applied to solve the equations (Eq. (13)) and (Eq. (14)), and then, the linearized system

$$\begin{aligned} N(\varepsilon_0, \kappa_0) + \frac{\partial N(\varepsilon_0, \kappa_0)}{\partial \varepsilon}(\varepsilon - \varepsilon_0) + \frac{\partial N(\varepsilon_0, \kappa_0)}{\partial \kappa}(\kappa - \kappa_0) &= 0, \\ M(\varepsilon_0, \kappa_0) + \frac{\partial M(\varepsilon_0, \kappa_0)}{\partial \varepsilon}(\varepsilon - \varepsilon_0) + \frac{\partial M(\varepsilon_0, \kappa_0)}{\partial \kappa}(\kappa - \kappa_0) &= M(x) \end{aligned}$$

can be solved for unknown values

$$\kappa = F(\varepsilon_0, \kappa_0, M(x)), \quad \varepsilon = G(\varepsilon_0, \kappa_0), \tag{16}$$

where ε_0 and κ_0 are the initial approximations, and $M(x)$ is determined from (Eq. (7)).

As the initial approximation at small values of the load P, the solutions obtained for the linear constitutive equations (Eq. (15)) were used. Since the computation is performed with a relatively small increment of P, in case of big values of P, one can use the computation results acquired at a previous step as the initial approximation for current step.

Having determined the change of median surface curvature from the equations (Eq. (13)) and (Eq. (14))

$$\kappa(x) = \begin{cases} \kappa_1(x), & \text{at } x \in [0, l/2), \\ \kappa_2(x), & \text{at } x \in [l/2, l], \end{cases}$$

one can write down a differential equation to determine the beam bend. For that purpose, the bend function is expressed as follows:

$$w(x) = \begin{cases} w_1(x), & \text{at } x \in [0, l/2), \\ w_2(x), & \text{at } x \in [l/2, l]. \end{cases}$$

Using the equation (Eq. (9)) and the beam's fixing conditions, a system of equations can be derived:

$$\begin{aligned} &\frac{d^2 w_1}{dx^2} = -\kappa_1, \quad \frac{d^2 w_2}{dx^2} = -\kappa_2, \\ &w_1(0) = w_2(l) = 0, \quad w_1(l/2) = w_2(l/2), \\ &\frac{dw_1(l/2)}{dx} = \frac{dw_2(l/2)}{dx}. \end{aligned}$$

The solution of these equations can be obtained using the methods of solving boundary-value problems for systems of ordinary differential equations. For that purpose, the modified collocation and least-residuals method [19–21] were applied.

Numerical analysis of deformation processes in VSE-1212 polymer matrix and VKU-28 structured carbon fiber is based on approximation of stress-strain curves and the three-point bending model.

Further, three different specimens with the geometrical sizes $l \times 2h \times b$ are considered:

1. specimen 1—VSE-1212 polymer matrix, 75 × 4.78 × 10.05 mm,
2. specimen 2—VKU-28 structured carbon fiber (the specimen was cut out along the reinforcement), 90 × 3.45 × 9.85 mm;
3. specimen 3—VKU-28 structured carbon fiber (the specimen was cut out perpendicular to the reinforcement), 90 × 3.40 × 9.95 mm.

In **Figure 2**, one can see the simulation results for beam three-point bending, obtained through different approaches to approximation of the constitutive equations, and their comparison with the experimental data.

Applying the linear dependencies to tension and compression has not resulted in adequate approximation even for 30% of curve. Using more complex than quadratic approximation laws

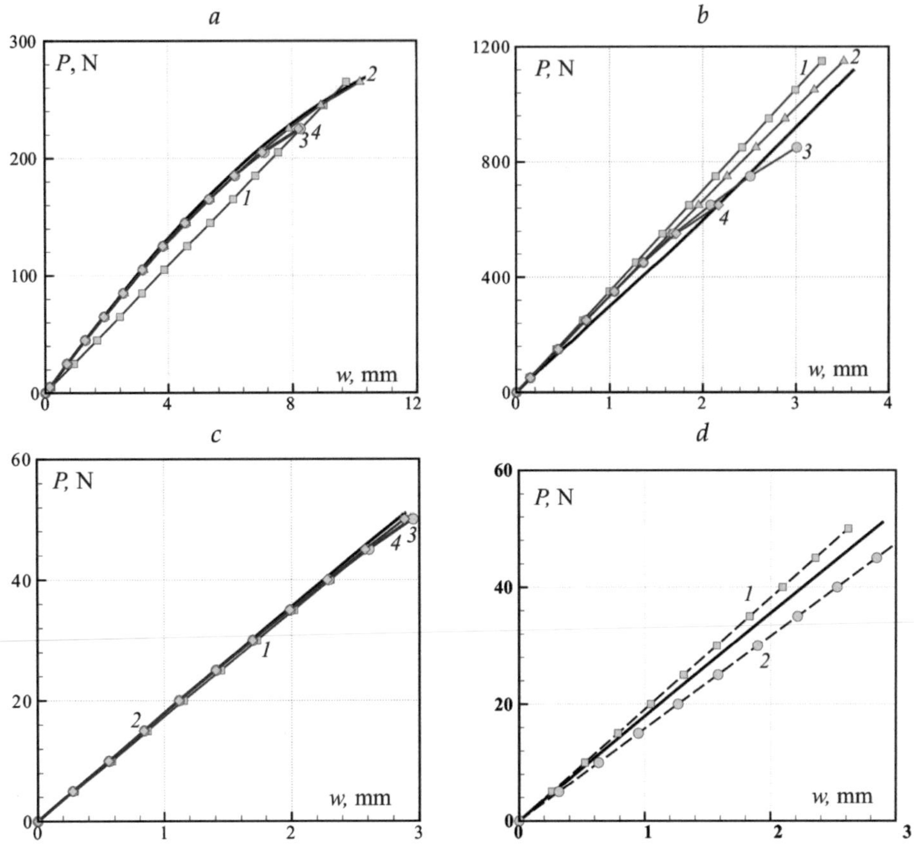

Figure 2. Experimental (solid curves) and dependencies of beam-deflection and load obtained in simulation: linear approximation (1); quadratic approximation by a polynomial of the second degree (2); cubic approximation (3); linear and power-law approximation (4); a–c are specimens 1 – 3 respectively; d—the solution to a three-point bending problem without account for the different strength and stiffness behavior in tension (curve 1) and compression (curve 2). The solid line shows the results of mechanical tests.

at first led to a significant deviation from the experimental curve and then to divergence of the Newton method iteration process. This is explained by the fact that in tension tests, due to specimens' fragility, the strain range for the polymer matrix specimens was limited to 2%, while in the bending tests, the strains in tension zone reached 4–5%.

Thus, to solve the bending problem, the tension curve was extrapolated into the domain of high strains. The extrapolations obtained using a polynomial of the third degree, and by linear and power-law function reached the maximum too quickly and then started to decrease, which is against the physics behind the deformation process. A similar effect was observed when calculating the bending of the carbon-fiber specimens cut out along direction of reinforcement filler.

The calculations using quadratic approximation and extrapolation of tension curves and approximation of compression curves within a short (up to 6%) segment have turned out to be best for qualitative and quantitative description of the nonlinear character of VSE-1212 polymer matrix bending. In the case of the specimen cut out perpendicular to direction of its reinforcement, all the approximations have shown the results close to experiment. At the same time for the test with the maximum load, the best option has still been application of quadratic approximations.

Taking different strength and stiffness behavior in tension and compression into account has an essential effect. As it was demonstrated earlier, the tension tests of VKU-28 specimens produced nonlinear stress-strain curves, while the difference of characteristics between tension and compression reached 5–7% for the longitudinal reinforcements and 12–15%—for the transverse ones (see **Tables 1** and **2**).

For the polymer matrix this difference exceeded 15% (see **Tables 3** and **4**).

Approximation type	**Approximation coefficients**			**MSD**
	$a_1 \cdot 10^{-9}$	$a_2 \cdot 10^{-9}$	a_3	
Tension of VKU-28 CFRP, $\varepsilon \in [0; 0.015]$				
A1	160.8			1.4e − 2
A2	144.9	1.44e + 3		5.7e − 4
A3	144.0	1.66e + 3	−1.14e + 13	4.3e − 4
A4	143.0	8.87e + 2	1.87	4.1e − 4
Compression of VKU-28 CFRP, $\varepsilon \in [0; 0.0018]$				
A1	155.4			5.8e − 3
A2	160.2	−3.33e + 3		3.4e − 3
A3	155.9	4.31e + 3	−3.00e + 15	2.9e − 3
A4	157.7	−5.81e + 8	3.98	3.0e − 3

Table 1. Approximation coefficients for stress-strain curves of VKU-28 carbon fiber specimens reinforced in longitudinal direction and mean square deviation (MSD) of $f(x)$.

Approximation type	**Approximation coefficients**			**MSD**
	$a_1 \cdot 10^{-9}$	$a_2 \cdot 10^{-9}$	a_3	
Tension of VKU-28 CFRP, $\varepsilon \in [0; 0.0076]$				
A1	7.37			1.1e − 2
A2	7.89	−9.22e + 1		6.4e − 4
A3	7.87	−8.97e + 1	2.54e + 11	4.3e − 4
A4	7.82	−2.23e + 2	2.20	4.8e − 4
Compression of VKU-28 CFRP, $\varepsilon \in [0; 0.0034]$				
A1	8.90			6.7e − 3
A2	9.21	−1.20e + 2		4.1e − 3
A3	8.96	1.16e + 2	−5.09e + 13	3.7e − 3
A4	9.04	−5.16e + 6	3.95	3.7e − 3

Table 2. Approximation coefficients for stress-strain curves of carbon fiber specimens reinforced in transverse direction and mean square deviation (MSD) of $f(x)$.

Approximation type	Approximation coefficients			MSD
	$a_1 \cdot 10^{-9}$	$a_2 \cdot 10^{-9}$	a_3	
Strain of VSE-1212 polymer matrix (Constant cross section) $\varepsilon \in [0; 0.018]$				
A1	3.30	—	—	2.9e − 2
A2	3.90	−4.38e + 1	—	1.5e − 3
A3	3.83	−3.17e + 1	−4.94e + 11	6.7e − 4
A4	3.80	−1.05e + 2	2.25	6.6e − 4
Strain of VSE-1212 polymer matrix (Variable cross section)$\varepsilon \in [0; 0.018]$				
A1	3.33			2.7e − 2
A2	3.89	−4.02e + 1		1.8e − 3
A3	3.80	−2.48e + 1	−6.30e + 2	4.0e − 4
A4	3.77	−1.40e + 2	2.35	2.7e − 4

Table 3. Approximation coefficients for tension curves of VSE-1212 polymer matrix, $\varepsilon \in [0; 0.018]$ and mean square deviation (MSD) of $f(x)$.

Approximation type	Approximation coefficients			MSD
	$a_1 \cdot 10^{-9}$	$a_2 \cdot 10^{-9}$	a_3	
Compression of VSE-1212 polymer matrix (Constant cross section), $\varepsilon \in [0; 0.28]$				
A1	0.77			2.3e − 1
A2	1.60	−3.97		8.2e − 2
A3	2.36	−1.29e + 1	2.37e + 10	1.4e − 2
A4	−5.71	5.49	0.90	3.8e − 2
Compression of VSE-1212 polymer matrix (Variable cross section), $\varepsilon \in [0; 0.28]$				
A1	0.69			3.5e − 1
A2	1.69	−5.22		1.4e − 1
A3	2.71	−1.84e + 1	3.84e + 1	3.6e − 2
A4	−2.07	1.72	0.72	4.8e − 2
Compression of VSE-1212 polymer matrix $\varepsilon \in [0; 0.06]$ (Variable cross section, shortened test area)				
A1	2.10			7.2e − 2
A2	3.05	−2.12e + 1		4.2e − 3
A3	3.18	−2.85e + 1	9.13e + 1	1.1e − 3
A4	3.31	−1.24e + 1	1.75	1.9e − 3

Table 4. Approximation coefficients for compression curves of VSE-1212 polymer matrix and mean square deviation (MSD) of $f(x)$.

Tables 1–4 show approximation results for the above-presented stress-strain curves by different functions at different intervals:

1. by the linear approximation $\sigma = a_1\varepsilon$ (**A1**),
2. by the polynomial of the second degree $\sigma = a_1\varepsilon + a_2\varepsilon^2$ (**A2**),
3. by the polynomial of the third degree $\sigma = a_1\varepsilon + a_2\varepsilon^2 + a_3\varepsilon^3$ (**A3**),
4. by a combination of linear and power-law functions $\sigma = a_1\varepsilon + a_2\varepsilon^{a_3}$ (**A4**).

However, if bending tests have been performed to determine an elasticity module of CFRPs, different strength and stiffness behavior in tension and compression is compensated and one obtains some averaged characteristic.

It is useful to consider the effect of the way for determining and setting of the mechanical characteristics on modeling of three-point bending of the carbon-fiber beam cutout perpendicular to its reinforcements. **Figure 2d** shows the solutions obtained while using a linear approximation of the constitutive equations with equal elastic moduli for tension and compression: for curve 1 the modulus was obtained from tension experiments, for curve 2—from compression ones (see **Table 2**).

As one can see the calculated linear results without account for the different strength and stiffness behavior in tension and compression have differed from the results of mechanical tests (the solid curve) by more than 15%.

Most of the real CFRP structures under day-to-day service conditions bear complex loads that result in formation of tension, compression and bending zones as well as their combinations in the structures. Applying the traditional methods for determination of material characteristics in combination with linear deformation models (in particular those that do not account for the different strength and stiffness behavior in tension and compression) for calculation of such structures, one risks to distort the deformation and stress pattern significantly, which, in its turn, results in either underestimation or overestimation of the structure's strength and rigidity. Keeping in mind that carbon fibers are used for manufacturing of high-duty structures, their computation demands different strength and stiffness behavior in tension and compression to be taken into account.

4. Numerical analysis and design of pressure vessels

Composite overwrapped pressure vessels (COPV) are used in the rocket and spacecraft making industry due to their high strength and lightweight. Consisting of a thin, nonstructural liner wrapped with a structural fiber composite COPV are produced to hold the inner pressure of tens and hundreds atmospheres. COPV have been one of the most actual and perspective directions of research, supported especially by NASA [22, 23].

Designing of a highly reliable and efficient COPV requires a technology for analysis of its deformation behavior and strength assessment. This technology should allow one to obtain target COPV parameters through changing vessel's geometry, structural and mechanical material parameters while keeping its useful load.

Application of combination mathematical modeling and numerical optimization makes it possible to reduce the cost and the duration of identifying the best parameters for a COPV. However, this approach is characterized by a number of hurdles. Overcoming these hurdles determines the success of an optimum designing of such structures.

So far, there have been two main approaches in optimization of composite structures: analytical and numerical ones.

In the first approach, the problems are solved basing on their simplified statement, for example using the momentless (membrane) shell theory and the netting model of composite material (CM) [24–27]. The obtained results may be far from reality; however, they are of value for testing of numerical optimization methods.

Application of the numerical approach in designing, on the other hand, produces a number of challenges that must be overcome, for example, lack of reliable methods for global optimization; nonconvexity and nonlinearity of constraint functions; ill-conditioned boundary value problems; different scaling of optimization criteria represents just some of the obstacles that prevent from reliable optimization of COPV.

Numerical analysis is usually a computation-intensive process and takes considerable time. One way to solve this problem is approximation of the objective function using different approaches, such as response surface method [28] and neural network [29]. Some kinds of numerical analyses use a small number of design variables, functions and/or corresponding set of their discrete values (analytical geometry parametrization [30], finite set of feasible winding angles [31]).

Another way is reasonable simplification of the elasticity problem statement, for example by using the membrane theory or other shell theories [30, 32, 33], that leaves the question of results validity. This is the approach we have applied in our study. For validation, we have used the Timoshenko [34] and Andreev-Nemirovskii [35] shell theories, accounting transverse shears with different degrees of accuracy.

Of course, it should be taken into account that the computed solutions are not optimum in the strict mathematical sense. However, these solutions could provide the considerable economy of the weight while keeping the required strength, and, therefore, they have high engineering value.

4.1. The problem statement and the mathematical models

Let us consider a multilayer composite pressure vessel at a state of equilibrium under equidistributed inner pressure. We need to determine the parameters of structure and CM meeting the following requirements:

$$V \geq V_0, \quad P \geq P_0, \quad M \leq M_0, \tag{17}$$

where V is the volume of the vessel, P is inner pressure and M is the vessel's mass and they are constrained by some preset values V_0, P_0, M_0.

We define the optimization problems the following way: to find extremum of one functional from (Eq. (17)) under other constraints.

The structural optimization problem statement includes selection of objective functional, formulation of constitutive equations and constraints on performance and design variables.

The mathematical models describing the vessel's state are based on the following assumptions:

1. the vessel is a multilayer thin-walled structure;
2. the vessel's layers can have different mechanical characteristics;
3. the reinforced layer's material is quasi-homogeneous;
4. the vessel's main loading is high inner pressure.

These assumptions allow us to reduce dimension of the corresponding mathematical problem and to build the mathematical vessel's models based on the different theories of multilayer nonisotropic shells.

Let us consider the vessel as a shell rigidly compressed on the edge. Taking into account a symmetry plane in the middle of the vessel, it is enough to calculate and design only its one half. The type of loading and boundary conditions allows considering the axisymmetric problem statement.

The shell is set by rotation of the generatrix $r = r(\theta)$ around axis $0y$ (**Figure 3**) where r is the current point of the shell radius, θ is the angle between the normal to the shell surface and the spin axis changing within $[\theta_0; \theta_1]$.

The Kirchhoff-Love shell theory [36] (KLST) and the improved Timoshenko [34] (TiST) and Andreev-Nemirovskii [35] (ANST) theories are used to solve the direct calculation problems of multilayer composite vessels, to analyze their behavior and to verify optimization problem solutions. The full systems of equations were described in the paper [17].

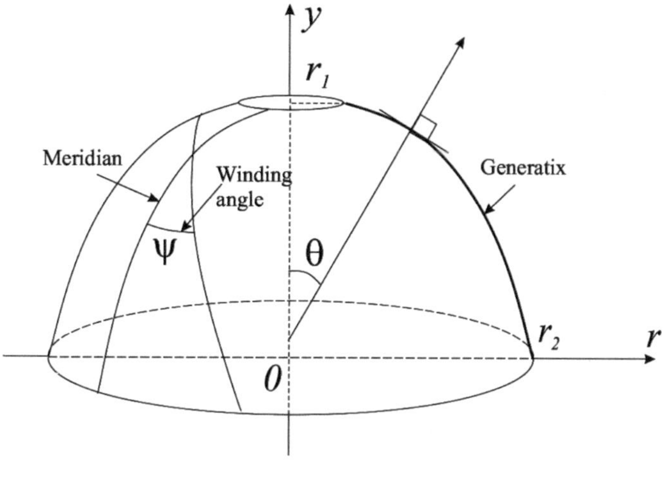

Figure 3. Shell of rotation geometry.

Relations between stresses and strains are described by the structural models [18]. The main idea of these models is that CM parameters are calculated through matrix and fibers mechanical parameters, fibers volume content and winding angles. The stress-strain state of matrix and fibers is evaluated through stresses and strains of the composite shell. A failure criterion is applied for every component of CM. Here we use the Mises criterion to determine the first stage of failure.

The objective function whose minimum is required is the minimum mass:

$$M = 2\pi \int_{\theta_0}^{\theta_1} rR_1 h d\theta [\rho_m (1 - \omega_r) + \rho_r \omega_r] \to \min, \tag{18}$$

where ρ_m, ρ_r are the densities of matrix and reinforcing fibers, ω_r is the volume content of reinforcement.

We chose the following design functions: the curvature radius $R_1(\theta)$ to define the generatrix; the thickness of the shell $h(\theta)$; the reinforcement angle $\psi(\theta)$ (**Figure 3**).

The solution has to satisfy the constraints on the shell's inner volume:

$$\pi \int_{\theta_0}^{\theta_1} r^2 R_1 \sin\theta d\theta = V_0 \tag{19}$$

and the strength requirement:

$$\max\{bs_r, bs_m\} \le 1, \tag{20}$$

where bs_r, bs_m are the normalized von Mises stresses in the matrix and fibers [1]. Note that the factor of safety is widely used while solving engineering problems. It can be considered by correction of the right-hand side of the inequality (Eq.(20)).

We used the following constraints on the design functions:

$$0 \le \psi \le 90, \quad h_0^* \le h \le h_1^*, \quad R_0^* \le R_1 \le R_1^*. \tag{21}$$

The method of the continuous geodesic winding has been widely used in the manufacturing of composite shells of revolutions. In this case the winding angles are defined by the Clairaut's formula:

$$r \sin\psi(r) = C, \tag{22}$$

where C—the constant is defined, as a rule, from the condition at the equator of the shell. The thickness equation is

$$h(r) = h_R \frac{R \cos\psi_R}{r \cos\psi(r)}, \tag{23}$$

which has the singularity at the edge where the winding angle has to be equal to 90°. The formula (Eq. (23)) is applied into practice at $r \geq r_0 + r_\omega$, where r_ω is equal to the width of the reinforcement tape. As a result, the equation determining the vessel's thickness takes the form:

$$h(r) = \begin{cases} h_R \dfrac{R \cos \psi_R}{r_\omega \cos \psi(r_0 + r_\omega)}, & r \leq r_0 + r_\omega; \\ h_R \dfrac{R \cos \psi_R}{r \cos \psi(r)}, & r \geq r_0 + r_\omega. \end{cases} \tag{24}$$

We did not consider the problem of fibers slippage. The main goal of the study was to demonstrate the potentials of using CM.

4.2. Direct problems: analysis of the shell theories

Estimation of composite vessel stress-strain state using offered models leads to the solution of boundary value problems for rigid systems of differential equations. These problems are ill-conditioned, and their solutions have pronounced character of thin boundary layers. Numerical analysis was performed by the spline collocation and discrete orthogonalization methods, realized in the COLSYS [37] and GMDO [38] software. These computing tools have proved to be effective in numerical solving of wide range of problems of composite shell mechanics [1].

We investigated the vessel's deformations by computing its stress-strain state based on the different shell theories. The vessel's shape was a part of a toroid: $R_1 = 2.46\,\text{m}$, $\theta_0 = 0.108°$, $\theta_1 = 90°$ (the computed half), $r(\theta_0) = 0.04$ m. The carbon composite parameters were: $E_m = 3 \cdot 10^9\,\text{Pa}$, $\nu_m = 0.34$, $E_r = 300 \cdot 10^9\,\text{Pa}$, $\nu_r = 0.3$, $\omega_r = 0.55$, $V_0 = 350$ liters where E_m, E_r are the Young's modulus of the matrix and fibers, ν_m, ν_r—their Poisson's ratio.

Figure 4 shows the stress-strain state characteristics of the vessel with the thickness $h = 0.6\,\text{cm}$, reinforced in the circumferential direction ($\psi = 90°$) under the load of 170 atm. On the left, the

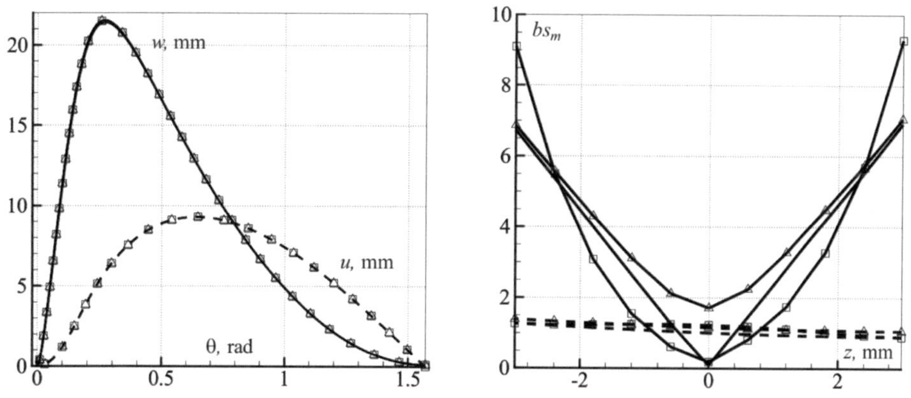

Figure 4. The stress-strain state characteristics of the composite vessel computed using different shell theories. Longitudinal displacement u—dashed curves; deflection w—solid curves. The curves without symbols correspond to KLST simulations, the curves marked with Δ—to those using TiST, and □—to ANST.

displacements of the reference surface along the generatrix $u_1(r)$ (dashed curves) and the normal displacement of these surfaces $w(r)$ (solid curves) are shown. On the right is the distribution of normalized von Mises stress (nVMS) along the thickness in the matrix $bs_m(r)$. The solid curves correspond to a slice at the shell edge, the dashed curves — to a slice at $\theta = 0.1$.

It is easy to see that the basic kinematic characteristics coincide both qualitatively and quantitatively. Small differences are observed only for the stresses and deformations near the compressed edge. The maximum results and qualitative difference were obtained for ANST. This is due to accounting for the transverse shears by nonlinear distribution in a thickness of a shell. Earlier it was shown [1] that ANST's-based results were the closest to the ones of 3D elastic theory in most cases.

The winding angle's influence on the COPV performance was investigated using parametric analysis. Dependence of the maximum nVMS in the matrix bs_m (dashed curves) and the fibers bs_r (dash-dotted curves), and the maximum size of the displacement vector $\| \vec{v} \|$ (solid curves) are shown in **Figure 5**.

The calculated values are very close in the area of their minima (**Figure 5** left side). The graphs of kinematic function $\|v\|$ coincide qualitatively. Some noticeable quantitative differences are revealed only for KLST's results.

The range $\psi \in (42; 45)$ corresponds to the zones of minimum values (**Figure 5** right side), which practically coincide ($\min_\psi bs_m \approx 0.65$, $\min_\psi bs_r \approx 1.05$, $\min_\psi \|v\| \approx 5 \cdot 10^{-3}$ m), as well as the angles, where these values are obtained ($\psi \approx 43.2^\circ$ for bs_m and bs_r, $\psi \approx 43.8^\circ$ for $\|v\|$).

It was revealed that the winding angles of minimum stresses values were almost insensitive to the thickness variation. The change of h from 0.6 to 1.6 cm corresponded to the angle's change about 0.2°.

Additionally, we investigated stress–strain state of the vessel (the thickness $h = 0.6\,\text{cm}$, the winding angles at $\psi = \pm 43.2$), when nVMS in the matrix and the fibers were near their minimum (**Figure 6**). The adopted notation is the same as in **Figure 4**.

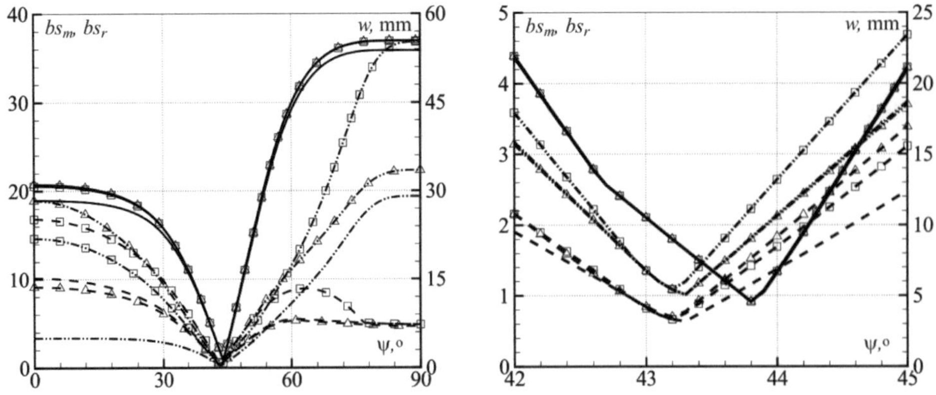

Figure 5. The winding angle's influence on the composite vessel stress-strain state. KLST's results are drawn without marks, TiST — with symbols Δ, ANST — with □.

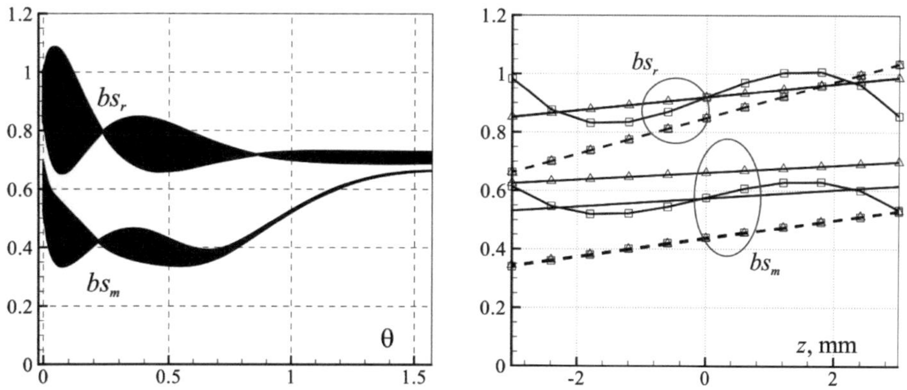

Figure 6. The stress-strain state of the vessel ($\psi = \pm 43.2$), computed using the three shell theories.

Again the difference is visible only in a very small region near the edge, but now this difference is small enough to be neglected. Moreover, the displacement values of the reference surface, the efforts and the moments completely coincide for all the theories.

All the theories (KLST, TiST, ANST) provided similar estimated characteristics of stress-strain state. This vessel was characterized not only by essential decrease of the maximal nVMS in the matrix and the fibers, but also by their uniform distribution along the generatrix. At the same time, the values of bending moments significantly reduced bringing vessel's stress-strain state close to momentless.

The performed analysis showed that the optimization problem can be solved using rather simple shell theories (KLST, TiST). These theories are characterized by lower computational complexity of corresponding boundary value problem if compared to ANST. It takes from 10 to 20 times less resources.

One can see that the winding angle as a design parameter gives an opportunity to increase the vessel's strength significantly. The difference between the "best" and "worst" designs can reach 20–35 times comparing their nVMS in the matrix and fibers. The "worst" designs have the winding angle close to 90°. In this case are considerable transverse shears near the compressed edge, and the loading is redistributed to a rather weak matrix while the fibers remain unloaded.

4.3. Inverse problems: optimization of the vessel

Inverse problems involve not only numerical methods for fast and reliable solving of direct boundary value problems, but also require numerical optimization methods for finding design parameters.

Here we considered conditional optimization problem, including direct constraints on design functions and trajectory constraints on the solution imposed at the end of the interval. The sequential unconstrained optimization is one of the most widespread approaches to solution of such problems. The main idea of the method is terminal functional convolution and multiple solutions of one-criterion problem using different optimization methods [39]. In our study, the modified Lagrange function was used for the convolution.

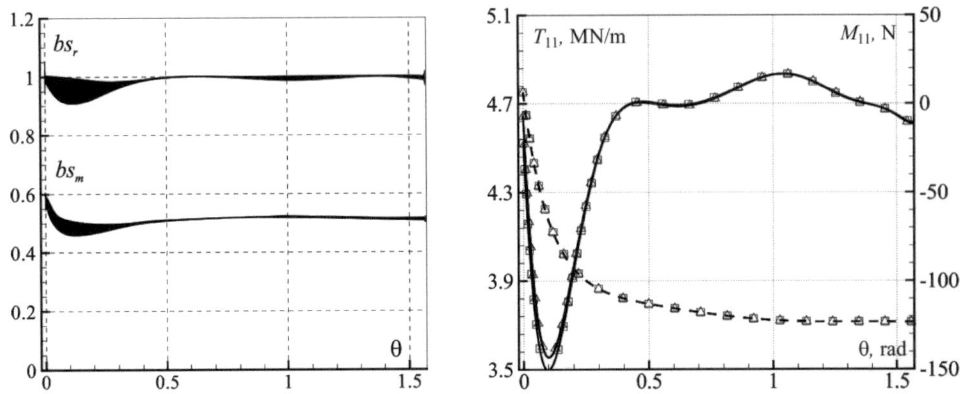

Figure 7. The stress-strain state characteristics of the vessel with the optimized design functions based on the three shell theories. Longitudinal force T_{11}—dashed curves; bending moment M_{11}—solid curves.

Hence we sought for solution of a nonconvex problem of finite-dimensional optimization [40] by discretization of design functions. The methods realized in the OPTCON-A software [41] were used to get the corresponding solution.

The considered design with the continuous geodesic winding has been one of COPV widely used in practice [42, 43].

Important additional design characteristic is its "adaptability in manufacturing." For example, the 5–10 times difference of thickness along the meridian would become a serious obstacle for vessels manufacturing. Thus, designs of nearly minimum mass possessing good properties and satisfying to the given technological constraints could be of great value than optimum without them.

According to Amelina et al. [44], the design with the geodesic continuous winding has the thickness ratio about 10 and large gradient near the edge.

We verified the solutions of optimization problem by substituting the obtained design parameters into the direct problem. In [44] shown that all three theories yielded close results (**Figure 7**). The difference is noticeable only for ANST in narrow zones (less than 1% of all area of calculation) at the edges, where non-linear accounting for transverse shear gives difference of about 5%. At the same time, the estimated efforts and bending moments are very close for all the theories, and the bending moments are very small.

Thus, it is possible to use the simplest shell theory to solve such optimization problem and the estimation of stress-strain state will be close to those obtained using more complex theories.

5. Conclusions

- Mathematical models for nonlinear flexural deformation of CFRPs and polymer matrices with account for their different strength and stiffness behavior in tension and compression have been built. A satisfactory match with the results of mechanical tests has been obtained. The study has proved that the nonlinear properties of polymer matrices and carbon fibers should be taken into account when calculating and designing real structures.

- The technology of optimization of COPV has been developed. It makes possible to obtain high pressure vessel designs that not only meet such requirements as minimum mass, preset volume and strength, but also possess a number of additional valuable engineering characteristics including stress-strain state close to momentless and almost equally stressed fibers.
- Nonconstant design parameters, such as thickness, winding angles and curvature radius of composite shell give the possibility for additional reduction of COPV mass while keeping its strength. The solutions of the optimization problem have been verified by solving the direct problems with obtained design parameters using the classical and improved shell theories.
- The study has demonstrated acceptability and convenience of using simple mathematical models based on Kirchhoff—Love and Timoshenko shell theories for numerical solving optimization problems.

Acknowledgements

The study was supported by project 18-13-00392 of Russian Science Foundation.

Author details

Sergey Golushko[1,2]*

*Address all correspondence to: s.k.golushko@gmail.com

1 Novosibirsk State University, Novosibirsk, Russia

2 Institute of Computational Technologies SB RAS, Novosibirsk, Russia

References

[1] Golushko S, Nemirovskii Yu. Direct and inverse problems of mechanics of composite plates and shells. Moscow: Fizmatlit; 2008. 432 p. (in Russian)

[2] Boyce M, Arruda E. An experimental and analytical investigation of the large strain compressive and tensile response of glassy polymers. Polymer Engineering & Science. 2004;**30**(20):1288-1298. DOI: 10.1002/pen.760302005

[3] Lagace P. Nonlinear stress-strain behavior of graphite/epoxy laminates. AIAA Journal. 1985;**23**(10):1583-1589. DOI: 10.2514/3.9127

[4] Sendeckyj G, Richardson M, Pappas J. Fracture behavior of Thornel 300/5208 graphite/epoxy laminate – Part I: Unnotched laminates. Composite Reliability, American Society for Testing and Materials. 1973;**STP580**:528-546

[5] Vasiliev V, Morozov E. Mechanics and analysis of composite materials. Elsevier. 2001 424 p

[6] Timoshenko S. Strength of Materials. Part 1, Elementary Theory and Problems. New York: Van Nostrand; 1955 359 p

[7] Ambartsumyan S. Theory of materials heteroresistant to tension and compression. Moscow: Nauka; 1982. 317 p. (in Russian)

[8] Ambartsumyan S. Strength of materials heteroresistant to tension and compression. Yerevan: RAU; 2004. 187 p. (in Russian)

[9] Jones R. Mechanics of composite materials with different moduli in tension and compression. Final scientific report AFOSR-TR-78-1400. 1978. 129 p

[10] Ambartsumian S, Gnuni V. The flexion of non-linear elastic beam taking into consideration the material resistance to the tension and compression in different ways. NAS RA Reports. Mechanics. 2005;**1**. http://elib.sci.am/2005_1/07/07r.htm (in Russian)

[11] Daniel I, Albot J. Fabrication, testing and analysis of composite sandwich beams. Composites Science and Technology. 2000;**60**(12–13):2455-2463. DOI: 10.1016/S0266-3538(00)00039-7

[12] Mostafa A, Shankar K, Morozov E. Experimental theoretical and numerical investigation of the flexural behaviour of the composite sandwich panels with PVC foam core. Applied Composite Materials. 2014;**21**:661-675. DOI: 10.1007/s10443-013-9361-4

[13] Mostafa A, Shankar K, Morozov E. Independent analytical technique for analysis of the flexural behavior of the composite sandwich panels incorporated with shear keys concept. Materials and Structures. 2014;**48**(8):2455-2474. DOI: 10.1617/s11527-014-0331-6

[14] Amelina E, Golushko S, Erasov V, Idimeshev S, Nemirovskii Y, Semisalov B, Yurchenko A, Yakovlev N. Nonlinear deformation of carbon fiber reinforced plastics: Experiment, model, and simulation. Computational Technologies. 2015;**20**(5):27-51 (in Russian)

[15] Bolotin V, Novitchkov Y. The mechanics of multilayered constructions. Moscow: Mashinostroeniye; 1980. 371 p. (in Russian)

[16] Nemirovskii Yu, Reznikov B. Strength of elements of composite structures. Novosibirsk SB RAS: Nauka; 1986. 166 p. (in Russian)

[17] Golushko S. Direct and inverse problems in the mechanics of composite plates and shells. In: Krause E, Shokin Yu, Resch M, Shokina N, editors. Computational science and high performance computing, Notes on Numerical Fluid Mechanics and Multidisciplinary Design. Vol.88. Springer; 2005. p. 205-227. ISBN 3-540-24120-5

[18] Golushko K, Golushko S, Yurchenko A. On modeling of mechanical properties of fibrous composites. In: Krause E, Shokin Yu, Resch M, Kroner D, Shokina N, editors. Computational

science and high performance computing IV, Notes on Numerical Fluid Mechanics and Multidisciplinary Design. Vol. 115. Springer; 2011. p. 107–120. DOI: 10.1007/978-3-642-17770-5

[19] Golushko S, Idimeshev S, Shapeev V. Application of collocations and least residuals method to problems of the isotropic plates theory. Computational Technologies. 2013; **18**(6):31-43 (in Russian)

[20] Golushko S, Idimeshev S, Shapeev V. Development and application of collocations and least residuals method to the solution of problems in mechanics of anisotropic laminated plates. Computational Technologies. 2014;**19**(5):24-36 (in Russian)

[21] Shapeev V, Belyaev V, Golushko S, Idimeshev S. New possibilities and applications of the least squares collocation method. EPJ Web Conferences. 2018;**173**:01012. DOI: 10.1051/epjconf/201817301012

[22] Beeson H, Davis D, Ross W, Tapphorn R. Composite overwrapped pressure vessels. NASA/TP—2002–210769; 2002

[23] Thesken J, Murthy P, Phoenix S, Greene N, Palko J, Eldridge J, Sutter J, Saulsberry R, Beeson H. A Theoretical investigation of composite overwrapped pressure vessel (COPV). Mechanics applied to NASA full scale tests. NASA/TM—2009–215684; 2009

[24] Stadler W, Krishnan V. Natural structural shapes for shells of revolution in the membrane theory of shells. Structural Optimization. 1989;**1**:19-27

[25] Banichuk N. Optimization of axisymmetric membrane shells. Applied Mathematics and Mechanics. 2007;**1**(4):578-586 (in Russian)

[26] Obraztsov I, Vasiliev V, Bunakov V. Optimum Reinforcing of Composite Shells of Rotation. Moscow: Mashinostroenie; 1977 144 p. (in Russian)

[27] Vasiliev V, Krikanov A, Razin A. New generation of filament-wound composite pressure vessels for commercial applications. Composite Structures. 2003;**62**:449-459

[28] Vafaeesefat A. Optimization of composite pressure vessels with metal liner by adaptive response surface method. Journal of Mechanical Science and Technology. 2011;**25**(11): 2811-2816

[29] Papadrakakis M, Lagaros N. Soft computing methodologies for structural optimization. Applied Soft Computing. 2003;**3**(3):283-300

[30] Liang C, Chen H, Wang C. Optimum design of dome contour for filament-wound composite pressure vessels based on a shape factor. Composite Structures. 2002;**58**:469-482

[31] Kim C, Kang J, Hong C. Optimal design of filament wound structures under internal pressure based on the semi-geodesic path algorithm. Composite Structures. 2005;**67**:443-452

[32] Zu L, Koussios S, Beukers A. Shape optimization of filament wound articulated pressure vessels based on non-geodesic trajectories. Composite Structures. 2010;**92**:339-346

[33] Fukunaga H, Uemura M. Optimum design of helically wound composite pressure vessels. Composite Structures. 1983;**1**:31-49

[34] Grigorenko Y, Vasilenko A. Static Problem of Anisotropic Inhomogeneous Shells. Moscow: Nauka; 1992 332 p. (in Russain)

[35] Andreev A, Nemirovskii Yu. Multilayer anisotropic shells and plates: bending, stability, oscillation. Novosibirsk: Nauka; 2001. 288 p. (in Russian)

[36] Novozhilov V. Theory of Thin Shells. Leningrad: Sudpromgiz; 1951 431 p. (in Russian)

[37] Ascher U, Christiansen J, Russel R. Collocation software for boundary value ODEs. ACM Transactions on Mathematical Software. 1981;**7**(2):209-222

[38] Golushko S, Yurchenko A. Solution of boundary value problems in mechanics of composite plates and shells. Russian Journal of Numerical Analysis and Mathematical Modeling. 2010;**25**(1):27-55

[39] Bertsekas D. Constrained optimization and Lagrange multiplier method. Moscow: Radio; 1987. 400 p. (in Russian)

[40] Evtushenko Yu. Methods of solution of extremum problems and their application in systems of optimization. Moscow: Nauka; 1982. 432 p. (in Russian)

[41] Gornov A. Computational Technologies for Optimal Control Solution. Novosibirsk: Nauka; 2009 275 p. (in Russian)

[42] Lepikhin A, Moskvichev V, Chernyayev A, Pokhabov Yu, Khalimanovich V. Experimental evaluation of strength and tightness of metal–composite high-pressure vessels. Deformation and Rupture of Materials. 2015;**6**:30-36 (in Russian)

[43] Amelina E, Burov A, Golushko S, Doronin S, Lepikhin A, Moskvichev V, Yurchenko A. Experimental and computational estimation for strength of a metal composite pressure vessel. Computational Technologies. 2016;**21**(5):3-20 (in Russian)

[44] Amelina E, Golushko S, Yurchenko A. Analysis and design of hybrid pressure vessels. In: CEUR Workshop Proceedings; Vol. 1839. (Mathematical and Information Technologies MIT-2016); 28 August-5 September 2016; Vrnjacka Banja, Serbia; Budva, Montenegro; 2017. p. 244-257

Chapter 3

Optimization of Lay-Up Stacking for a Loaded-Carrying Slender Composite Beam

Sergey Shevtsov, Igor Zhilyaev,
Natalia Snezhina and WU Jiing-Kae

Additional information is available at the end of the chapter

http://dx.doi.org/10.5772/intechopen.76566

Abstract

Many aircraft composite structures experiencing the high operational loads must have the specified mechanical stiffness to prevent some structural failure due to the inadmissible deformations. Usually, such parts are manufactured using composites with orthotropic symmetry, which provides the best combination of structural rigidity, strength, and weight. In this chapter, we consider a cantilevered long tube-like composite structure with varied cross-section that is manufactured by winding of glass fiber unidirectional tape. The operational loads include the bending forces and the distributed torques. To reduce the total strain energy and peak von Mises stress, the search of the best lay-up scheme and its angles is performed. The wall thickness, lay-up scheme, and the total number of layers for each modeled design are assumed as unchanged along the tube, whereas its mechanical properties are considered as homogenized and dependent on the lamina properties and lay-up scheme only. The search of the pseudo-optimal design includes the analysis of all moduli angular distributions for each lay-up stacking. The better solutions are then studied by using the finite element model of the structure for three most critical load scenarios. The choice of the most preferred design is made by discarding the solutions with sharply degraded structural rigidity at least at one load scenario.

Keywords: structural optimization, aircraft composite structures, lay-up design

1. Introduction

The glass/carbon fiber composites are widely used in the design of various aircraft and rotorcraft components such as stringers, the spars of the rotor blades, the stiffened panels of the wings and stabilizers, which have predominantly a thin-walled geometry and are made of composite laminates that most often are adopted as orthotropic. The main requirements for these parts are the specified dynamic and endurance properties to provide the aerodynamic quality and reliability of the aircraft assembly at the different flight conditions with limited part weight [1]. Most of these parts are manufactured by laying-up or winding of the unidirectional lamina and curing in closed molds or autoclaves where the parts are exposed to the controlled pressure and temperature according to the predetermined process.

The rapid growth of composite applications in aircraft and rotorcraft industries, the very complex mechanical behavior of anisotropic composite materials require new structural strength, rigidity, modeling, optimization and failure prediction methods using virtual testing analysis method, which is a concept with several attributes and is to be understood as the simulation of aircraft structure [2, 3]. In the survey paper [4] the main milestones of development of the optimization and its applications in aircraft systems design, especially in the context of the composite structures, are considered. According to the concept, which is present in the considered paper, a full workflow process of the composite structure optimization consists of the following three phases:

- Free size or topology optimization
- Detailed design: ply-bundle sizing with ply-based FE (finite element) modeling
- Detailed design: ply stacking sequence optimization

During the second and third phases the composite plies with the different orientation are chosen and shuffled to determine the optimal stacking sequence for the given design optimization problem while satisfying some additional manufacturing constraints. As outlined in the article [5], such a composite structure optimization involved a large number of both the design variables and the objectives. As a result, designers must consider only a restricted set of parameters among the most influential. In the paper [5], the author uses the following parameters of optimized laminates: mass per unit area of an elementary plate, the compliance of elementary plate and the stiffness of a composite plate experienced by the uniform pressure. The stacking sequence, total number of plies and prepreg materials are accepted as the design variables, whereas the mass and the total number of layers are constrained. It was established that the ply orientation seems to have a significant influence on the most common mechanical criteria (mass, rigidity, etc.). Use of a more refined orientation set can improve the mechanical performance when accounting for inter-laminar stresses, or for laminates made of a few anisotropic plies. In most engineering applications, when there is a large number of plies, it is entirely warranted to restrict the layers' orientations to a limited number of predefined values. Similar technologies and appropriate soft tools are being developed and improved by leaders of aircraft industry such as Altair Engineering GMBH [6].

Because the dynamic behavior of the aircraft structures at the operating conditions being within the structure strength limits is very important, many investigations are limited by the study of composites at the elastic conditions. Such elastic phenomena were the main assumptions used at the modeling and were formulated in the early fundamental work [7]. These assumptions include: (1) The results obtained assume that the composite material may be modeled as a single-phase anisotropic homogeneous continuum. This assumption is adequate for the elastic constants, provided the main consideration is the interface, but is unlikely to apply to the strength properties. (2) There is a representative volume element whose mechanical properties are equal to the average properties of the particular composite. (3) The constituent materials are linearly elastic and homogeneous, but in general anisotropic. (4) Linearity of the stress-strain relations is assumed to hold for the composite materials.

Most investigations that have been implemented in this area use the finite element modeling for solve the forward structural mechanics problem combined with some optimization algorithms based on the soft computing concepts, for example, genetic algorithm [8–10], shuffled Frog-Leaping (SFL) algorithm [11] for lay-up sequence optimization of laminate composite structures. However, there are no theorems to prove that a soft algorithm ensures the achievement of a unique global optimum. Therefore, some researchers use analytical approaches to optimize the simplest composite systems [12] or empirical methods for more complex ones [13, 14].

Some works are devoted to the optimization of composite structures subjected to aeroelastic or hydraulic dynamic loading. The common part of such works is the subtask of simulating aeroelastic actions on a composite structure. In particular, the paper [15] examined phenomenon of bending-torsional coupling rigidity for achieving an optimal design of a composite wing with a maximum flutter speed. In order to ensure an effective way to achieve an optimal lay-up, two optimization approaches have been investigated. In the first approach, the flutter speed was set in the objective function directly. In the second approach, optimization was carried out to minimize an objective function containing the torsional and coupling rigidities rather than flutter speed. Kalavalapally and co-workers [16] solving a multidisciplinary optimization problem for a lightweight torpedo model subjected to underwater explosions observed that the composite torpedo model is stronger and lighter than the metallic design when subjected to an underwater explosions at a given standoff distance. The paper [17] presented the numerical method and results of the structural optimization of the mounting zones of wind turbine blades to diminish and flatten the stress distribution at the action of the extremely stressed wind load on the blade, which had the stiff carbon/epoxy composite skin and less stiff lightweight core body.

When composite structure with complex shape is studied, its CAD model needs to be used for the virtual aerodynamic testing by the means of FE tools. The main objective of the paper [18] is the optimization of wall thickness and lay-up sequence of shell-like cowling made of the transversely isotropic multilayered composite. The used approach assumed: (1) conversion of the CAD model of the cowling surface to the FE representation, then (2) wind tunnel testing simulation at the different orientation of airflow to find the most stressed mode of flight that uses at the formulation of the loading pressure conditions for the forward problem of structural mechanics. The optimization problem used the global strain energy calculated within the optimized shell as the objective, whereas the total weight of the optimized part was considered as the design constraints.

The important conclusion that can be made taking into account the surveyed works is the multiobjective nature of most composite optimization problems. Such a conclusion can be made on the basis of the present chapter, which is broken into several parts. First, we present a brief description of the optimized composite structure, which is the tube-like cantilever slender beam experiencing distributed bending and torsion forces. Then we determine the elastic properties of laminates used in the modeled tube. We start from the mechanical properties of reinforcing fibers and epoxy resin, and then we determine the properties of the unidirectional lamina and, finally, the laminate properties. Our optimization approach contains three sequential stages. The preliminary stage is based on the consideration of the angular distribution of all engineering constants of laminates. This analysis allows us to choose the small enough set of "candidate" lay-ups, which should be used at the modeling of the mechanical response of the studied structure at three different load scenarios. The next "candidates"—higher level "candidates"—are appointed for the final dynamic test, which includes applying full load to the preferred structures and gives us the possibility to make the expert decision about final choice of quasi-optimal structure. Last, we discuss some considerations influencing the final choice of the best lay-up parameters that are the design variables.

2. Modeled structure: geometry and operating conditions

The composite beam studied in this research is a slender tubular structure with optimizing prepreg lay-up sequence. Its cross-section is comprised of two straight stripes and two half-circles, each of them is made of eight laminating layers (see **Figure 1**), that is similar to those studied in [19]. The simulated structure is a greatly simplified design of D-like spar of the helicopter main rotor blade. Such spars are typically characterized by thickening of the cross section near the root, decreasing thickness of the airfoil and its twist as it approaches the end of the blade, without tapering of airfoil. The CAD model of the structure was built by using NX CAD (Siemens®) capabilities and converted into the Structural Mechanics Comsol Multiphysics environment.

In order to define the structural anisotropy of the laminate and the distributed external loads, the curvilinear coordinate systems $(\mathbf{e}_1, \mathbf{e}_2, \mathbf{e}_3)$ have been intended on the external surface of the tube. To do this, we used a diffusion method which solves Laplace's equation $\Delta U = 0$ and computes the vector field as $-\nabla U$. The boundary conditions were assigned as follows: on the internal surface $U_{\text{int}} = 1$, $U_{out} = 0$ on the outer surface, and $-\mathbf{n} \cdot \nabla U = 0$ on the end faces. The normal to the surface is assigned as $\mathbf{e}_1 = -\nabla U / |-\nabla U|$. The second component is the unit vector parallel to the beam longitudinal axis and third one $\mathbf{e}_3$ is normal to the plane $(\mathbf{e}_1, \mathbf{e}_2)$. The curvilinear coordinates are presented in **Figure 2**.

The root end face is fixed $(u, v, w) = 0$, where u, v, w are the displacements along the global axes (x, y, z), respectively. Three different external loads can be applied alternately or together:

$$F_{tors} = \begin{cases} 0 \cdot \mathbf{e}_1 \\ 0 \cdot \mathbf{e}_2 \\ 10^4 \cdot H(x-3, 1) \cdot \mathbf{e}_3 \end{cases} \tag{1}$$

$$F_{bend} = \begin{cases} 0 \cdot \boldsymbol{\xi} \\ 70 \cdot yload \cdot \boldsymbol{\eta} \\ 45 \cdot zload \cdot \boldsymbol{\zeta} \end{cases} \tag{2}$$

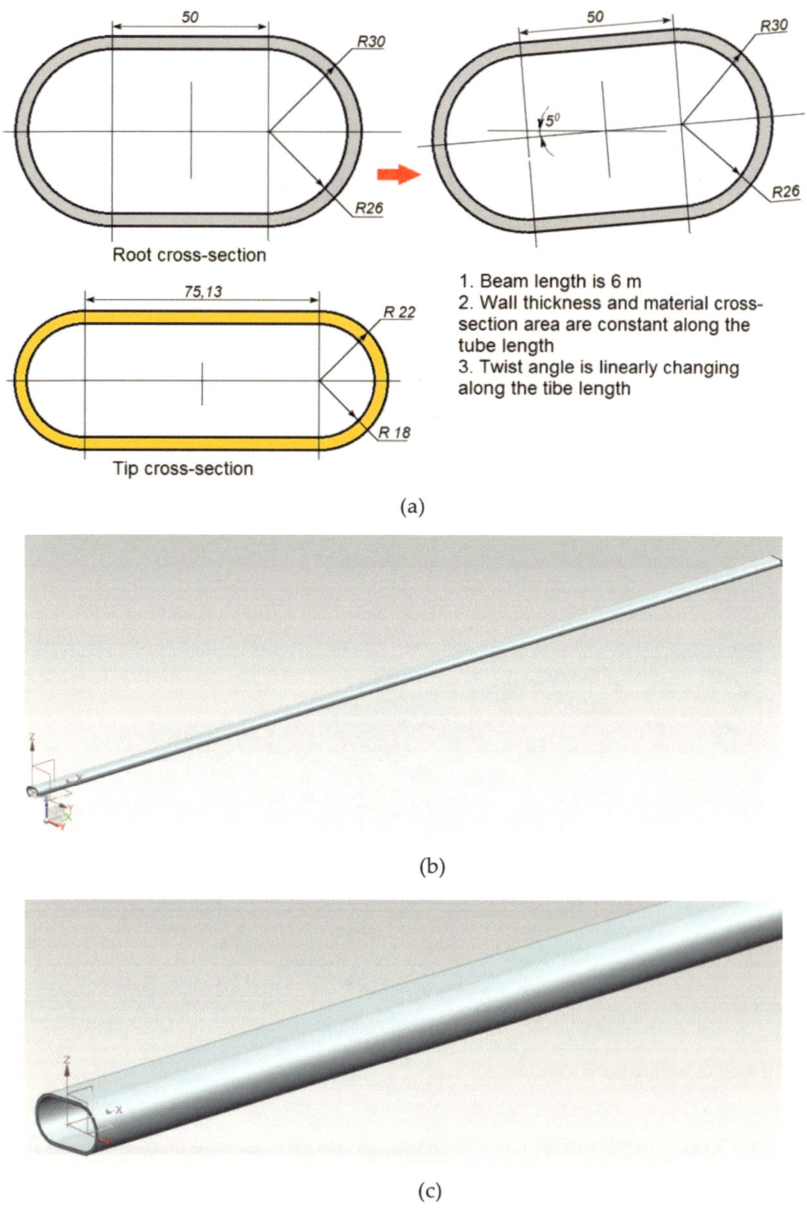

Figure 1. Sketch of the beam cross-sections (a), 3D view of the CAD model and magnifying view on the cross-section of optimizing structure.

In Eq. (1) the twisting load $H(x, \delta x)$ is the smoothed step-function with the coordinate of step x and smoothing width δx. The bending force F_{bend} is defined by the components in the global coordinate system whose unit vectors are (ξ, η, ζ) and by the parameters $yload, zload$, which can take alternatively or together, have the values of 0 and (or) one.

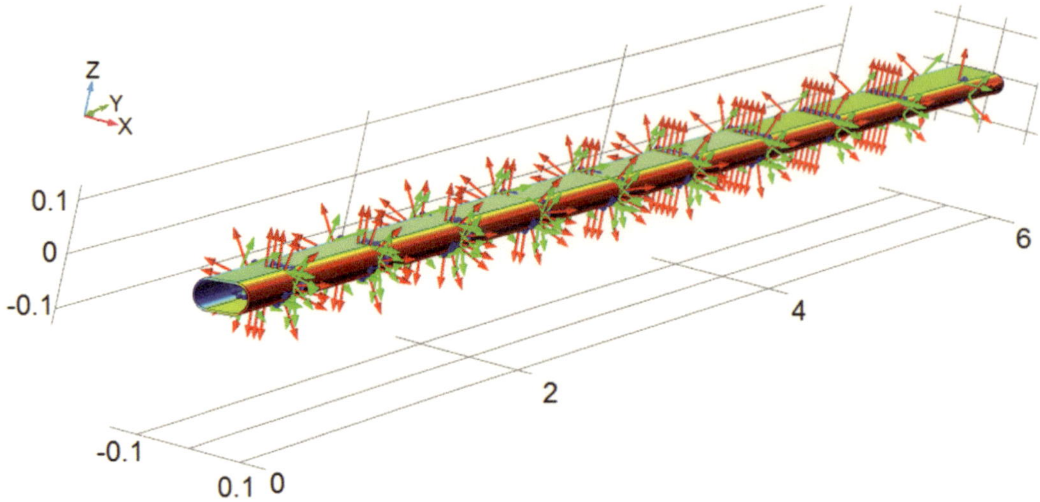

Figure 2. Curvilinear local coordinate systems on the beam surface.

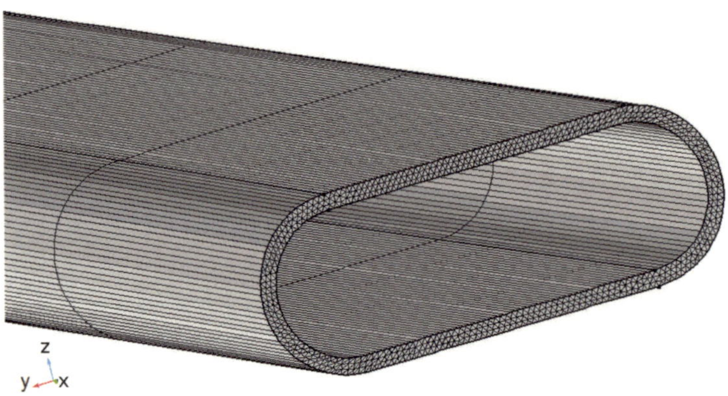

Figure 3. FE mesh used at the mechanics problem.

The FE mesh has been constructed for the given geometry as swept mesh (see **Figure 3**) consisting of 128,893 domain elements, 37,959 boundary elements and 1350 edge elements. The number of degrees of freedom solved for is 625,409.

3. Determination of lamina properties

All numerical and experimental studies have been implemented for glass fiber reinforced epoxy based prepreg VPS-7. Initially, the properties of the fibers and resin, which was accepted as isotropic, have been studied experimentally using testing machines Galdabini. The samples of resin and bundles of the fibers were experienced by tensile forces with strains

Components	Young module, GPa	Poisson ratio	Shear module, GPa
Glass fibers	87.5 (longitudinal)	$\nu_{12} = \nu_{13} = 0.255$	$G_{12} = G_{13} = 2.594$
Epoxy resin	2.55	0.364	0.94 (calculated)

Table 1. Mechanical properties of the composite constituents.

Elastic module	Determination method		
	Experiment	Mixing rule	FE modeling
E_1, MPa	**49,500 ± 2500**	**49,000**	**48,090**
E_2, MPa	6000 ± 500	5500	**8042**
ν_{12}	**0.315 ± 0.015**	**0.31**	**0.299**
ν_{23}	NA	NA	**0.459**
G_{23}, MPa	2600 ± 200	2000	**2758**
G_{12}, MPa	NA	NA	**2847**

Remark: Index 1 corresponds to the longitudinal ply orientation (along the fibers), whereas index 2 to the transversal ply orientation.

Table 2. Mechanical properties of lamina.

monitored by extensometers. During these experiments, the following elastic moduli have been obtained (see **Table 1**).

The properties of unidirectional lamina, which has transversely isotropic symmetry, were determined both numerically and experimentally for the samples with volume fraction $V_f = 0.534$. In order to determine the parameters inaccessible through the experimental tests, we used the mixing rule and FE modeling in Abaqus environment according to the recommendation of [20, 21]. The values all found properties are shown in **Table 2**, where bold numbers denote the most reliable values, which have been used in the follow-up investigation. These values have been used for determination of elastic moduli of laminates of chosen structures.

4. Determination of elastic moduli for laminates

In order to diminish the formation of the residual stress and strains, which can arise during cure we use only 8-layers symmetric balanced laminates in our investigation. All used schemes of lamina stacking are shown in **Figure 4** with their designations.

For each lay-up scheme the elastic moduli were determined independently by two methods: by the finite element method and on the base of the classical laminates theory. The last one proposes the following relationships for the elastic moduli of laminate.

$$\bar{E}_x = \frac{A_{11}A_{22} - A_{12}^2}{A_{22}H}; \quad \bar{E}_y = \frac{A_{11}A_{22} - A_{12}^2}{A_{11}H};$$
$$\bar{G}_{xy} = \frac{A_{66}}{H}; \qquad \bar{\nu}_{xy} = \frac{A_{12}}{A_{22}}. \tag{3}$$

where indices x, y correspond to the longitudinal and transversal directions of laminate, respectively, and the elements of $\{A\}$ matrix are calculated by the formula [22, 23].

$$A_{ij} = \sum_{k=1}^{N} \bar{Q}_{ijk}(z_k - z_{k-1}) \tag{4}$$

Where, $\bar{Q}_{ijk}$ are the i, j-th elements of the reduced stiffness matrix $\bar{Q}$ for the k-th layer, z_{k-1}, z_k are the bounds of the k-th layer and H is the laminate thickness.

The FE model for determining of elastic moduli simulated the experimental tests according to the standards ASTM D 3039–95 and ASTM D 5379–93 with the numerical methods for the

Figure 4. Used schemes of lamina stacking.

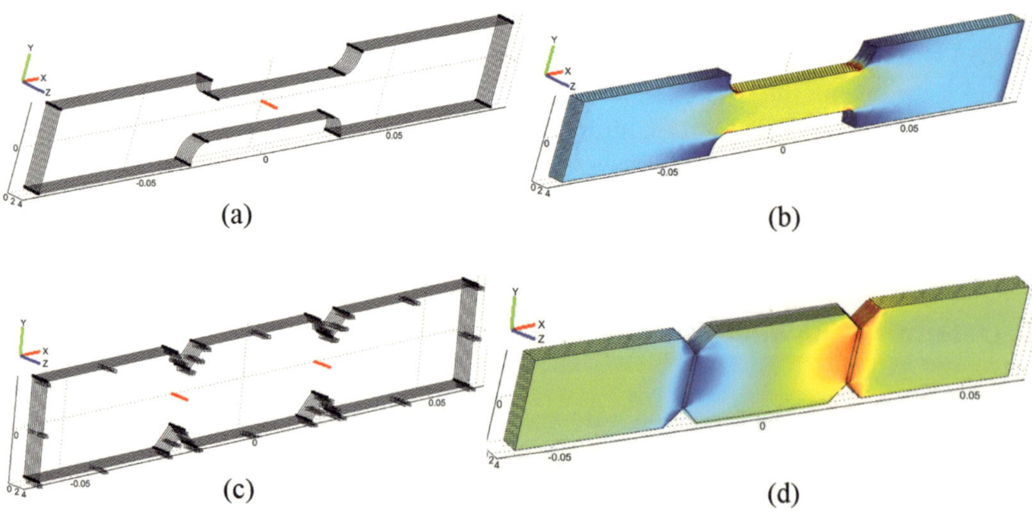

Figure 5. Geometry (a, c) and postprocessing results (b, d) of models simulating the mechanical tests ASTM D 3039–95 "Test Method for Tensile Properties of Polymer Matrix Composite Materials" (a, b) and ASTM D 5379–93 "Test Method for Shear Properties of Composite Materials by V-Notched Beam Method" (c, d).

accurate identification of the modules described in [24]. All modules to be determined in the numerical experiments are averaged by integration along the lines where the pure tensile or shear occurs (see **Figure 5**).

Geometry of the FE models consists of eight separated layers, each of which is characterized by its orientation of lamina. Numerical calculation of elastic modules has been implemented for all lay-ups (see **Figure 4**) with the step $\Delta\phi = 10^0$ of lamina orientation.

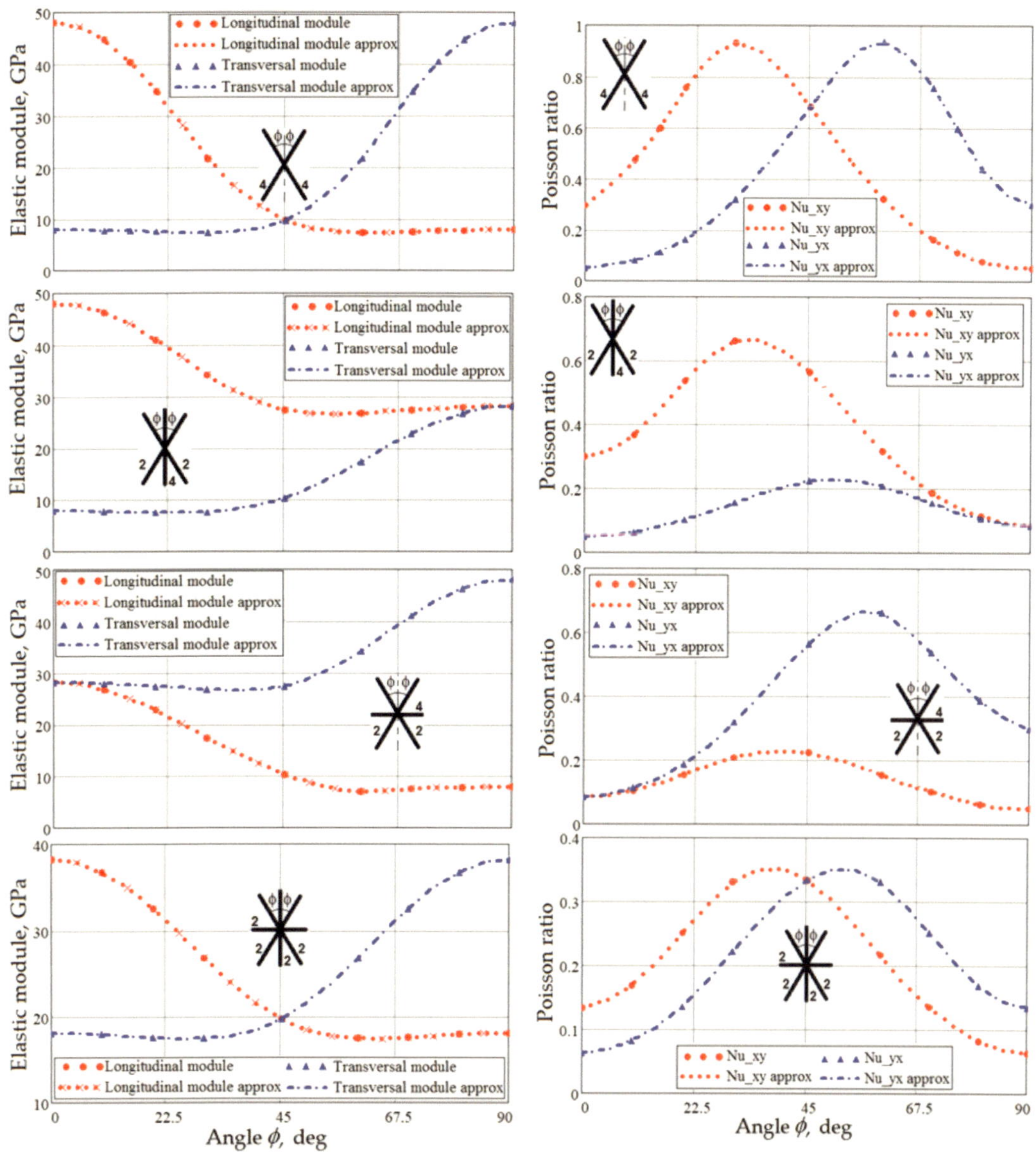

Figure 6. Some examples of longitudinal, transverse elastic moduli and Poisson ratios dependencies on the lay-up angles for the laminate stacking I, II, III, IV (see **Figure 4**).

Both moduli calculation methods have demonstrated a good correspondence, but FE method provides higher accuracy at the calculation of the Poisson ratios and in-plane shear module G_{xy}. Some examples of obtained results are shown in **Figures 6** and **7**. Each point on these plots at the chosen angle ϕ (or ϕ and ψ for the scheme V) represents the

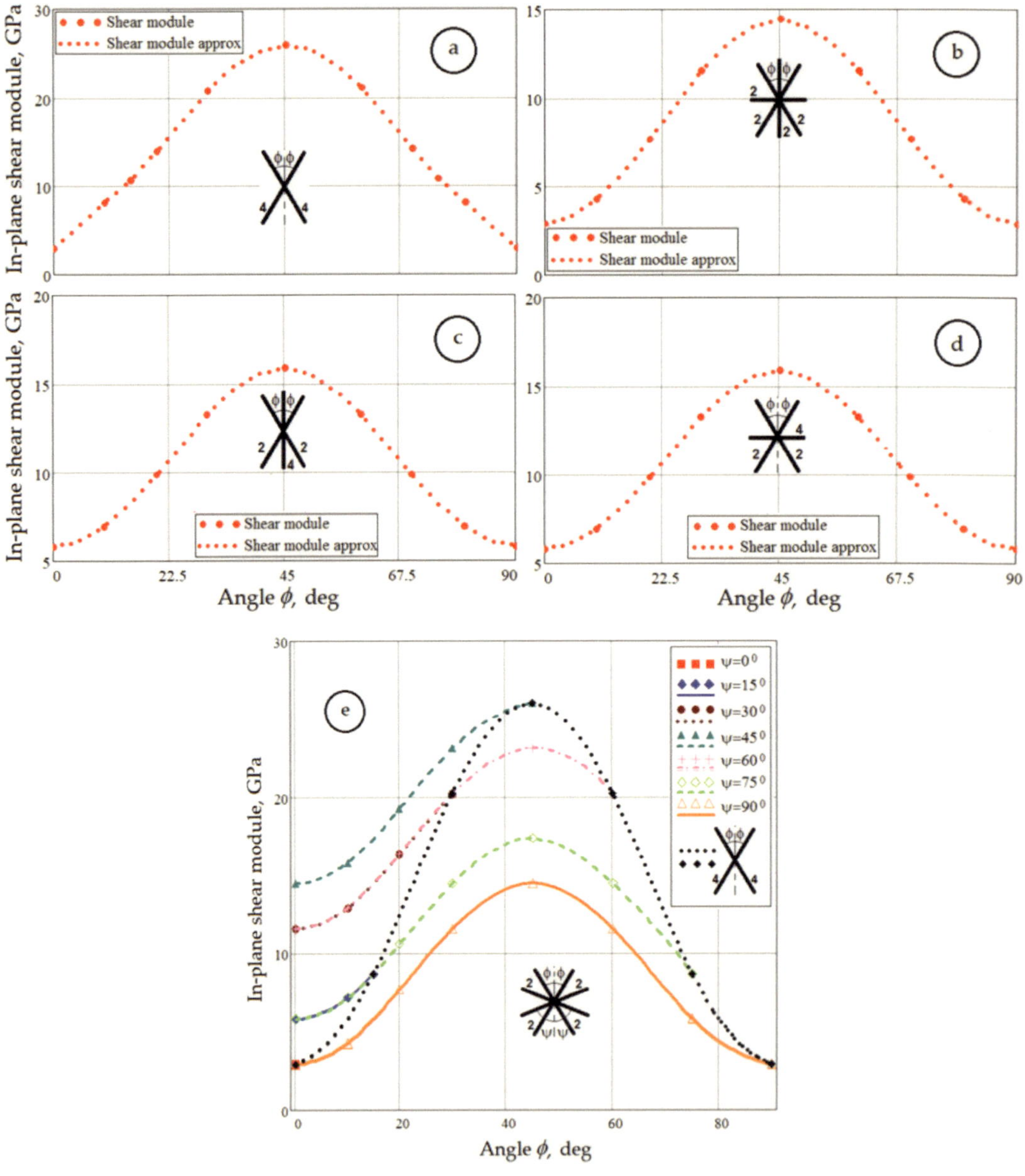

Figure 7. The in-plane shear module dependencies on the lay-up angles for the laminate stacking I, II, III, IV and V (see **Figure 4**).

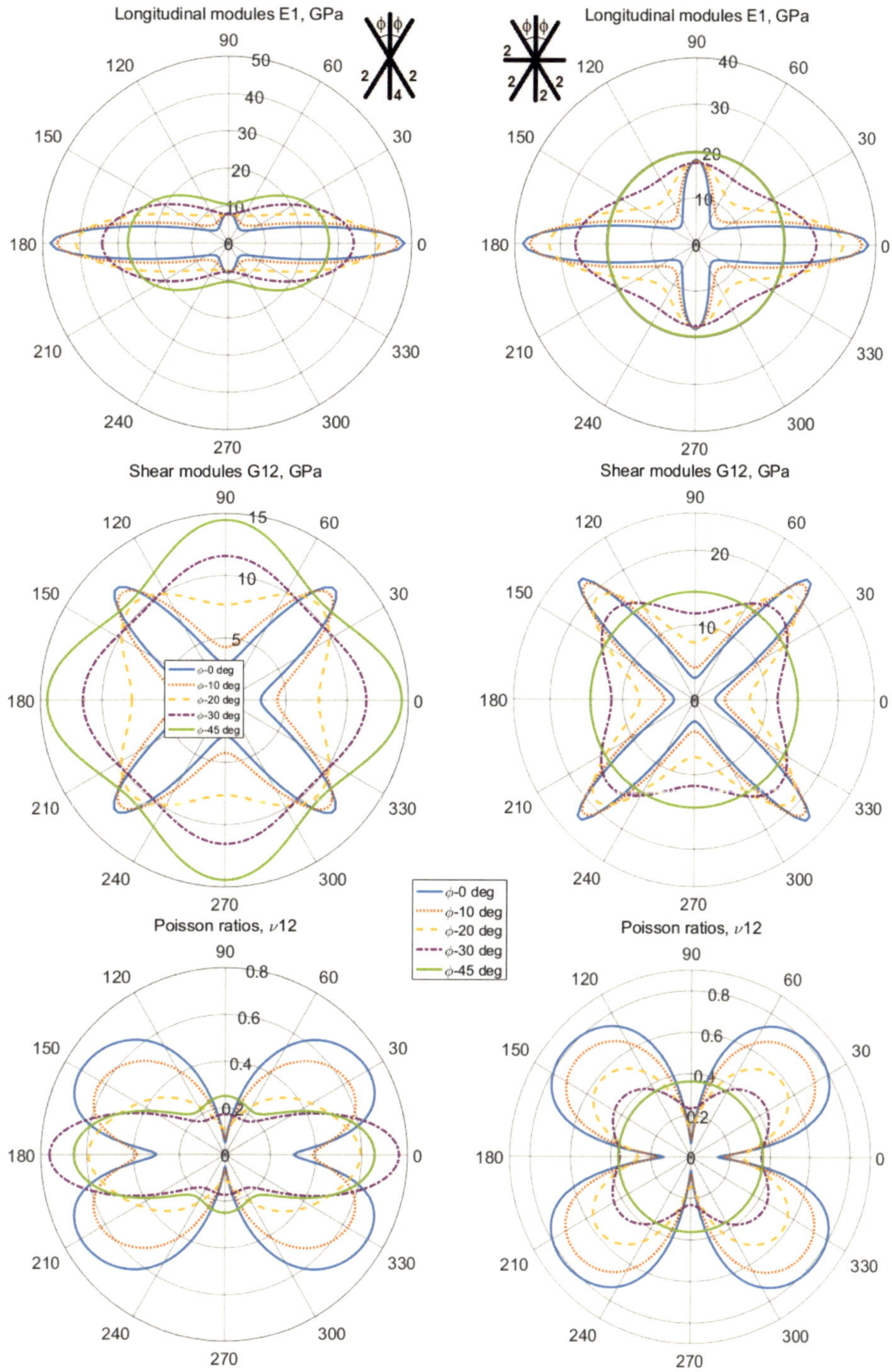

Figure 8. Angular dependencies of the longitudinal, shear modules and in-plane Poisson ratio for the lay-ups II (left) and IV (right) (see **Figure 4**).

values of effective elastic moduli for the laminate with given lay-up. Note that according to the homogenization hypothesis this laminate is considered as the solid body with uniform structure.

In the next stage of investigation, the angular dependencies of the effective moduli were investigated. In order to avoid the significant reduction of the structural rigidity in some directions, these dependencies should not have sharp drops, corresponding to these directions. Some examples of such angular dependencies are shown in **Figure 8**. These dependencies have been used for the preliminary selections of the "candidates" lay-ups, which have been further used to test them in mechanical testing of the optimized slender structure.

5. Mechanical testing of structure with different laminate lay-ups

The mechanical testing of the structure with different lay-ups and ply stacking angles has been implemented using FE model (see **Figures 1–3**) by using three abovementioned types of static bending and torsion loading. These loads were applied separately; for each load scenario the total strain energy E_{el}^{tot}, the maximum and averaged von Mises stress, two end deflections $\max(v_b)$, $\max(w_b)$ and torsion angles θ on the free face of the beam tip were calculated. At the solving of these subtasks a linear parametric solver was used. Results of these testing were used for the refinement of the "candidates," that is, select lay-ups that provide the minimum strain energy, bending and torsion deformations. Our study demonstrated a strong coupling of total strain energy of deformed structure with the values of maximum beam deflection (bending load cases) and maximum torsion angles (twisting load case). Moreover, two very interesting and important facts were established. First, we found that sensitivity of both maximum and averaged von Mises stress to the structural symmetry of orthotropic material is noticeably less comparing to the total elastic strain energy. Thereby we do not give here any von Mises dependencies. Second result is the practically independent response of the structure on the twisting load of the lay-ups II, III and IV. The reason for this is the practical identity of the shear modulus of these lay-up schemes at the same lamina angular orientation (see **Figure 7b–d**).

The calculated dependencies of maximum deflections and torsion angles of the loaded beams on the lay-up parameters are present in **Figure 9**. These results together with the angular dependencies of elastic moduli that are partially presented in **Figure 8** allowed to select eight "candidates" for further optimization. These candidates had to have acceptable values for the total strain energy and rigidity under all loading scenarios.

6. Pseudo-optimal choice of the preferred lay-ups

In the final stage of investigation, the optimizing structure made of the preferred lay-ups "candidates" has been loaded simultaneously by the systems of three forces: bending by the

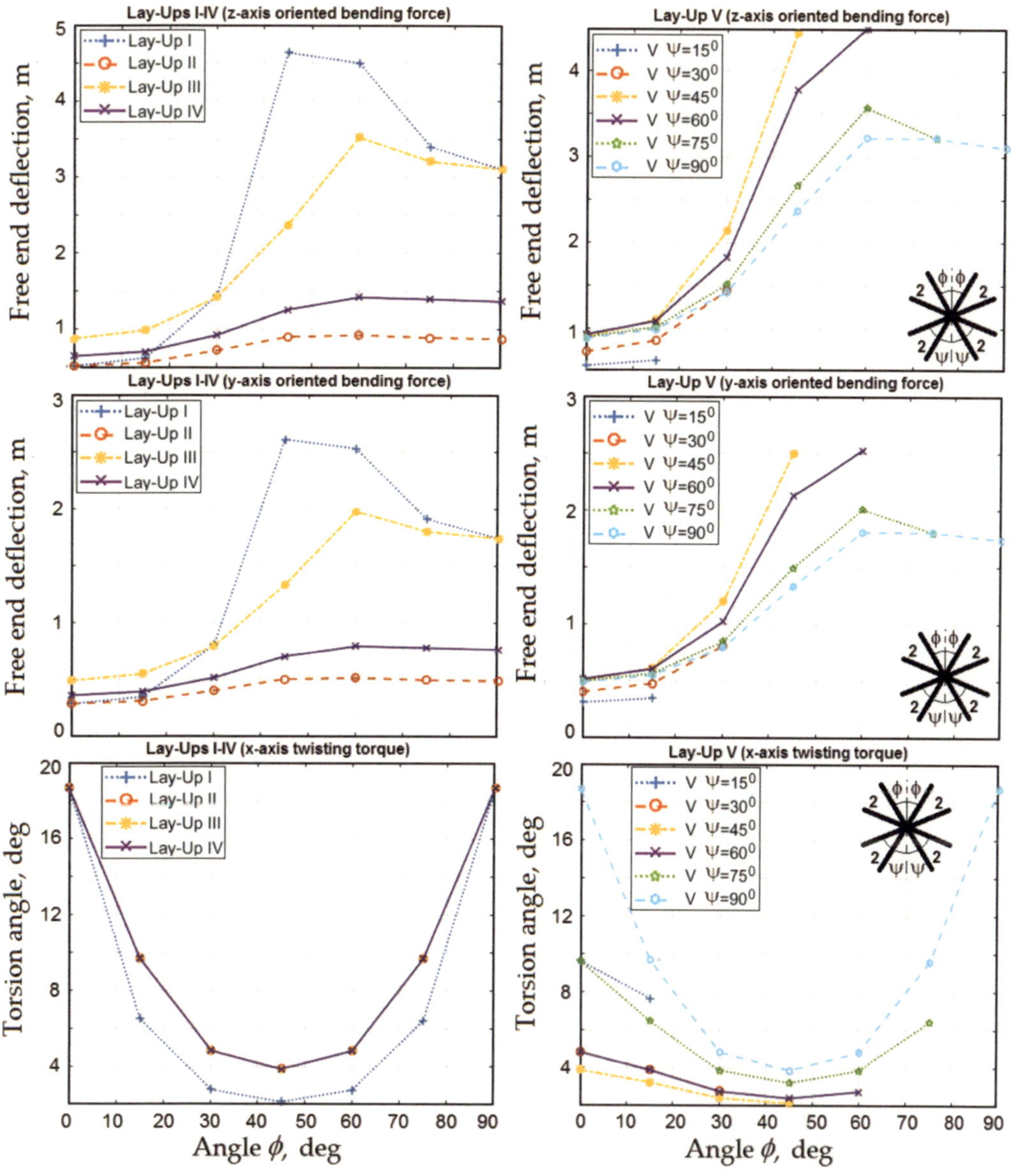

Figure 9. Dependencies of maximum deflections and torsion angles of the loaded beams.

distributed forces oriented along y and z axis and twisting torque applied to the external surface and given by Eqs. (1) and (2). The studied responses included total strain energy, the maximum bending deflections and objective, which was accepted in the normalized dimensionless form.

$$Obj = \theta/6 + \max(v_b) + \max(w_b)/1.5 \quad (5)$$

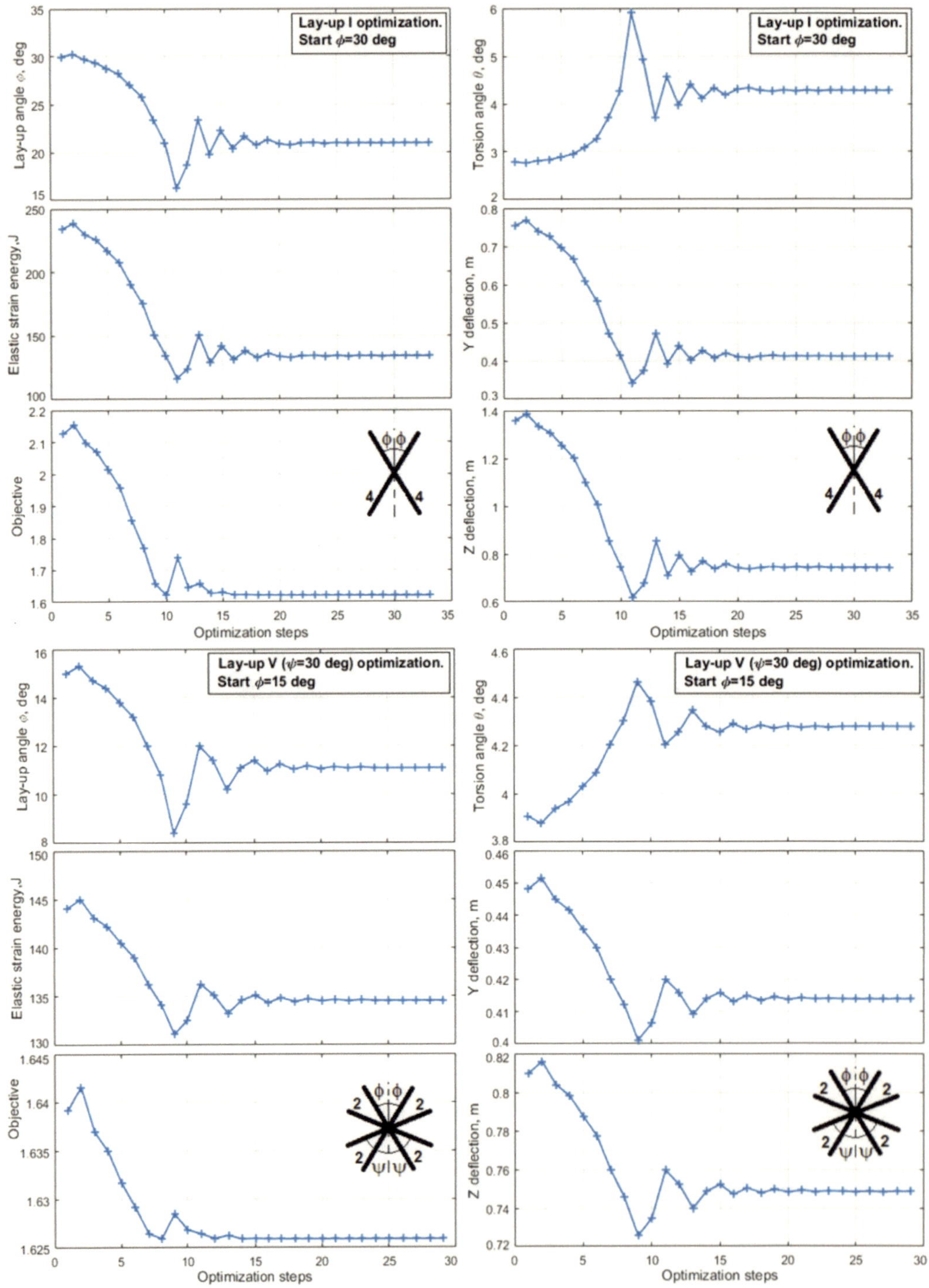

Figure 10. Evolution of lay-up angle, objective and mechanical properties of structure during optimization for two "candidate" lay-ups: (I $\phi_{in} = 30^{0}$) and (V $\psi_{in} = 30^{0}$; $\phi_{in} = 15^{0}$).

Where, θ is the torsion angle and v_b, w_b are the beam tip deflections along the y and z, respectively. Naturally, definition of such objective function and weights for its components has a significant element of arbitrariness. Therefore, the total strain energy, the beam deflections and torsion angle were also monitored during optimization process along with the objective function determined by Eq. (5).

For the definiteness of the optimization problem statement the following constraints were imposed.

$$\begin{array}{ll} E_{ei}^{tot} \le 160J \ ; & \theta \le 4.5^0 \ ; \\ \max(v_b) \le 0.46m \ ; & \max(w_b) \le 0.86m \end{array} \tag{6}$$

At the optimization workflow, the forward structural mechanics problem has been called by the built-in gradient-free Nelder-Mead optimizer. In order to start the iterative optimization procedures, we used such "candidates": (I $\phi_{in} = 30^0$); (II $\phi_{in} = 30^0$); (III $\phi_{in} = 30^0$); (IV $\phi_{in} = 30^0$); (V $\psi = 30^0; \phi_{in} = 15^0$); (V $\psi = 45^0; \phi_{in} = 15^0$); (V $\psi = 60^0; \phi_{in} = 15^0$) and (V $\psi = 75^0; \phi_{in} = 30^0$), which had

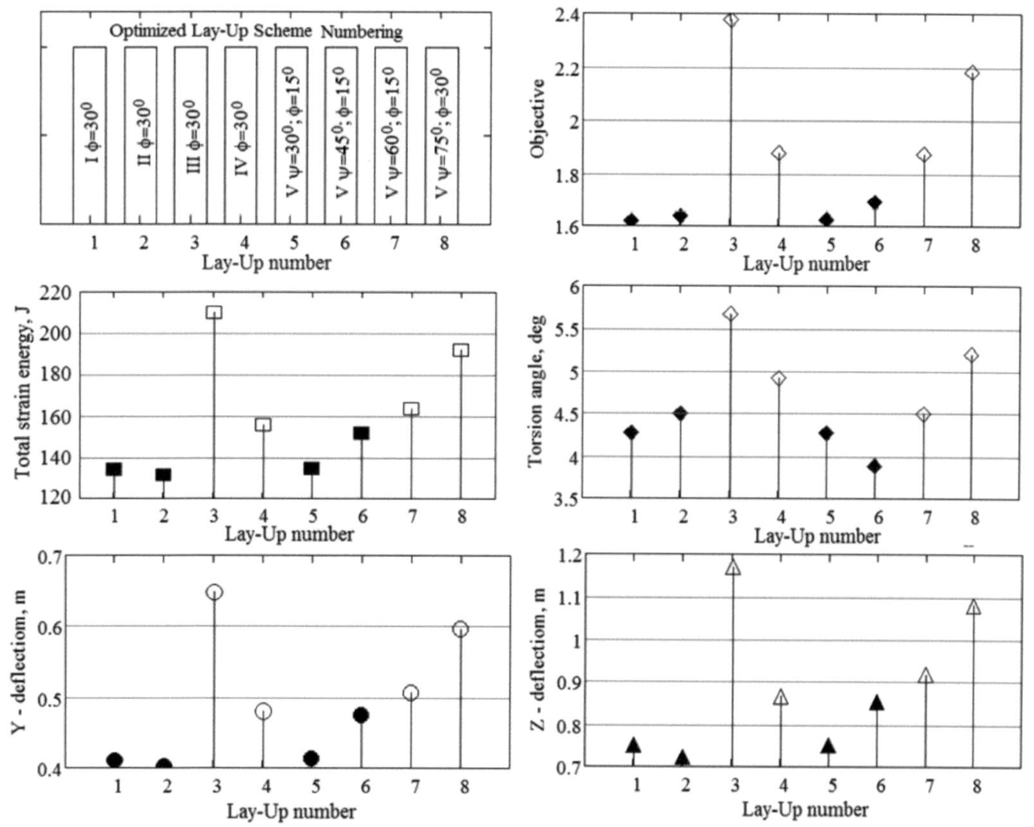

Figure 11. Comparison among the best lay-ups of the optimized structure (stems with the filled symbols correspond to the preferred lay-ups).

better results at the previous stage of the study (see **Figure 9**), where Roman numeral indicates the type of lay-up (see **Figure 4**) and the value of angle is its start value.

Figure 10 illustrates two examples of objective and other responses evolution during optimization process. These plots demonstrate very fast convergence to the quasi-optimal solutions for both optimized lay-up stacking. These solutions are not global optimum because they characterized by the different values of all responses.

The final features of all optimized "candidate" structures are present in **Figure 11**. All optimized lay-ups were ranked according to calculated responses. The optimized lay-ups (I $\phi_{in} = 30^0$) and (V $\psi = 30^0; \phi_{in} = 15^0$) provide minimal values of objective, but (II $\phi_{in} = 30^0$) provides the minimal total strain energy. Lay-up (V $\psi = 45^0; \phi_{in} = 15^0$) provides the greatest torsional stiffness, but its flexural rigidities are downscale. These considerations substantiate obligatoriness of multiobjective optimization at the design of load-carrying multilayered composite structures with orthotropic symmetry of the materials. The final decision can be made taking into account some constraints and requirements, for example, complexity of manufacturing, weight of ready structure, the natural vibration modes and eigenfrequencies and importance of a particular rigidity for the operability of the structure. An additional study of the strength of the composite layers according to the Tsai-Wu criterion [25] should be carried out at critical loads. Meanwhile, the used approach requires many tedious calculations, but allows us to obtain the visual substantial representations about chosen composite lay-ups for optimizing structures, and can be expanded to the composite structures with arbitrary number of arbitrarily oriented unidirectional layers.

7. Conclusion

This chapter studies a problem of lay-up optimization for a cantilevered long tube-like composite structure with varied cross-section that is manufactured by winding of glass fiber unidirectional tape. The optimized composite structure is the tube-like cantilever slender beam experiencing distributed bending and torsion forces. The multilayered composite material assumed and modeled as a single phase anisotropic elastic homogeneous continuum. We determine the elastic properties of laminates, which used in the modeled tube, taking as input data the mechanical properties of reinforcing fibers and epoxy resin to determine initially the elastic properties of the unidirectional lamina. For each accepted lay-up scheme and unidirectional prepreg orientation of the symmetric balanced laminate formation, the elastic moduli were determined independently by two methods: by the finite element method and on the base of the classical laminates theory.

The first stage of used optimization approach is based on the analysis of the angular distribution of all engineering constants of laminates. This analysis allows us to choose the small enough set of "candidate" lay-ups, which should be used at the modeling of the mechanical response of the studied structure at three different load scenarios. The higher level "candidates" were appointed for the final dynamic test, which includes applying full load to the selected structures and gives us the possibility to make the expert decision about final choice of quasi-optimal structure. The short discussion of the obtained results confirms necessity of multiobjective approach to the studied optimization problem, taking into account many requirements and constraints that allows to make the final choice of the best lay-up parameters.

Acknowledgements

This work is implemented in the framework of the state mission for the South Scientific Center of Russian Academy of Science, project No. AAAA-A16-116012610052-3 and it was supported by the Russian Foundation for the Basic Research (Grant 15-08-99849 and Russian-Taiwanese project 16-58-52013 MNT_a).

Author details

Sergey Shevtsov[1*], Igor Zhilyaev[1], Natalia Snezhina[2] and WU Jiing-Kae[3]

*Address all correspondence to: sergnshevtsov@gmail.com

1 South Center of Russian Academy, Rostov on Don, Russia

2 Don State Technical University, Rostov on Don, Russia

3 National Kaohsiung University of Science and Technology, Kaohsiung City, Taiwan R.O.C.

References

[1] Baker A, Dutton S, Kelly D. Composite Materials for Aircraft Structures. 2nd ed. Reston: AIAA Inc.; 2004. p. 597. ISBN 1-56347-540-5

[2] Ostergaard MG, Ibbotson AR, Le Roux O, Prior AM. Virtual testing of aircraft structures. CEAS Aeronautical Journal. 2011;**1**:83-103. DOI: 10.1007/s13272-011-0004-x

[3] Taylor JE. A formulation for optimal structural design with optimal materials. In: Rozvany GIN and Olhoff N, editors. Topology Optimization of Structures and Composite Continua. Dordrecht/Boston: Kluver Academic Publishers; 2000. pp. 49-55. ISBN 0-7923-6806-1

[4] Rao JS. Advances in aero structures. In: Proceedings of 12th International Conference on Vibration Problems (ICOVP-2015); December 14-17, 2015; Guwahati. p. 20

[5] Gillet A, Francescato P, Saffre P. Single- and multi-objective optimization of composite structures: The influence of design variables. Journal of Composite Materials. 2010;**44**(4):457-480. DOI: 10.1177/0021998309344931

[6] Optimization of Composite: Recent Advances and Application [Internet]. 2011. Available from: www.altairproductdesign.com [Accessed: February 01, 2018]

[7] Zhang WC, Evans KE. Numerical prediction of the mechanical properties of anisotropic composite materials. Computers & Structures. 1988;**29**(3):413-422

[8] Herencia JE, Hatfka RT, Weaver PM, Friswell MI. Lay-up optimization of composite stiffened panels using linear approximations in lamination space. AIAA Journal. 2008;**46**(9):2387-2391. DOI: 10.2514/1.36189

[9] Satheesh R, Naik GN, Ganguli R. Conservative design optimization of laminated composite structures using genetic algorithms and multiple failure criteria. Journal of Composite Materials. 2010;**44**(3):369-387. DOI: 10.1177/0021998309347579

[10] Gyan S, Ganguli R, Naik GN. Damage-tolerant design optimization of laminated composite structures using dispersion of ply angles by genetic algorithm. Journal of Reinforced Plastics and Composites. 2012;**31**(12):799-814. DOI: 10.1177/0731684412448104

[11] Mohan Rao AR, Lakshmi K. Optimal design of stiffened laminate composite cylinder using a hybrid SFL algorithm. Journal of Composite Materials. 2012;**46**(24):3031-3055. DOI: 10.1177/0021998311435674

[12] Oktem AS, Soares CG. Higher order theory based Fourier analysis of cross-ply plates and doubly curved panels. Journal of Composite Materials. 2011;**46**(21):2675-2694. DOI: 10.1177/0021998311431641

[13] Kaleemulla KM, Siddeswarappa B. Influence of fiber orientation on the in-plane mechanical properties of laminated hybrid polymer composites. Journal of Reinforced Plastics and Composites. 2010;**29**(12):1900-1914. DOI: 10.1177/0731684409340806

[14] Metiche S, Masmoudi R. Analysis and design procedures for the flexural behavior of glass fiber-reinforced polymer composite poles. Journal of Composite Materials. 2013;**47**(2):207-229. DOI: 10.1177/0021998312438721

[15] Guo SJ, Bannerjee JR, Cheung CW. The effect of laminate lay-up on the flutter speed of composite wings. Proceedings of the Institution of Mechanical Engineers, Part G: Journal of Aerospace Engineering. 2003;**217**(3):115-122. DOI: 10.1243/095441003322297225

[16] Kalavalapally R, Penmetsa R, Grandhi R. Multidisciplinary optimization of a lightweight torpedo structure subjected to an underwater explosion. Finite Elements in Analysis and Design. 2006;**43**:103-111. DOI: 10.1016/j.finel.2006.07.005

[17] Oganesyan P, Zhilyaev I, Shevtsov S, Wu JK. Optimized design of the wind Turbine's composite blade to flatten the stress distribution in the mounting areas. In: Dynybyl V, editor. The Latest Methods of Construction Design. Dordrecht/Heidelberg: Springer; 2016. pp. 335-341. DOI: 10.1007/978-3-319-22762-7

[18] Shevtsov S, Zhilyaev I, Oganesyan P, Axenov V. Optimization of wall thickness and lay-up for the shell-like composite structure loaded by non-uniform pressure field. AIP Conference Proceedings. 2017;**1798**:0200144-1-0200144-020014410. DOI: 10.1063/1.4972736

[19] Hodges DH. Nonlinear Composite Beam Theory. Reston: AIAA, Inc.; 2006. 303 p. DOI: 10.2514/4.866821

[20] Barbero EJ. Finite Element Analysis of Composite Materials Using Abaqus. Boca Raton: CRC Press; 2013. p. 406. ISBN 978-1-4665-1663-2

[21] Ha SK, Huang Y, Han HH, Jin KK. Micromechanics of failure for ultimate strength predictions of composite laminates. Journal of Composite Materials. 2010;**44**(20):2347-2361. DOI: 10.1177/0021998310372464

[22] Gibson RF. Principles of Composite Materials Mechanics. Detroit: McGraw-Hill, Inc.; 1994. p. 425. ISBN O-07-023451-5

[23] Voyiadjis GZ, Kattan PI. Mechanics of Composite Materials with MATLAB. Berlin/ Heidelberg: Springer; 2005. p. 336. ISBN:10 3-540-24353-4

[24] Chinchan L, Shevtsov S, Soloviev A, Huang JP. Mechanical testing of composites for aircraft structures: Standards, requirements and limitations. In: Parinov IA, Chang SH, editors. Advanced Materials. Switzerland: Springer; 2014. pp. 201-222. DOI: 10.1007/978-3-319-03749-3_17

[25] Tsai SW, Hahn HT. Introduction to Composite Materials. Westport: Technomic Publishung Co Inc; 1980. p 457. ISBN 0-87762-288-4

Chapter 4

Design Optimization and Higher Order FEA of Hat-Stiffened Aerospace Composite Structures

Bo Cheng Jin

Additional information is available at the end of the chapter

http://dx.doi.org/10.5772/intechopen.79488

Abstract

Sizing of hat-stiffened composite panels is challenging because of the broad design hyperspace in several geometric and material parameters available to the designer. Design tasks can be simplified if parameter sensitivity analysis is performed a priori and design data is made available in terms of a few important parameters. In this chapter, design sensitivity analysis is performed using finite element analysis (FEA) and analytical solution models. Manufacturing and experimental measurements of a hat-stiffened composite structure is performed to validate the FEA and idealized analytical solutions. This is an attempt to initiate a structural architecture methodology to speed the development and qualification of composite aircraft that will reduce design cost, increase the possibility of content reuse, and improve time-to-market. In particular, FEA results were compared with analytical solutions to develop a design methodology that will allow extensive reuse of parametric hat-stiffened panels in the design of composites structural components. This methodology is now widely utilized in developing a library of commonly used, built-in, composite structural elements in design of modern aircrafts. In this chapter, hat stiffened composite panels' geometric parameter sensitivity analysis work were parametrically investigated using finite element analysis (FEA), analytical solution models and experimental testing on manufactured parts in order to develop structural architectures that speed development and qualification of composite aircraft which has broad benefits in reducing cost, increasing content reuse and improving time-to-market. In particular, FEA results were compared with analytical solutions and a design methodology was developed to allow extensive reuse of parametric elements in structural design of composites and to achieve expedited design, verification, validation, and airworthiness certification and qualification. The goal of this work is to provide the aviation industry with the most up-to-date databases for the application of advanced composite materials incorporated into parametric models to eliminate redundancies in the current process. The work results include a correlated material database, an optimized model component library and a standardized way to design future complex composites structures, e.g. hat stiffened composites panels, with reliable and predictable quality and material weight/cost.

Keywords: design optimization, FEA, hat stiffeners, aerospace composite structures

1. Introduction

1.1. Background

In recent years the commercial aircraft industry is increasing their reliance on composite materials to produce lighter and more durable aircrafts. **Figure 1** shows a Boeing 787 aircraft contains 50% by weight of its materials as composites, which is about 32,000 kg of carbon-fiber-reinforced polymer (CFRP) [1].

The carbon fiber composites have a higher strength-to-weight ratio than traditional metal materials thus help making the aircraft lighter and to exceed the fuel efficiency target. Due to this important feature, the use of fiber reinforced composite laminates as primary structural components in these important large-scale and weight-critical applications has increased considerably (**Table 1**). Aircrafts with major composite parts including fuselage, wings, tail sections, doors and interior are presently being developed and gradually brought into service.

For better efficiency in terms of strength and weight-optimization, aerospace structures are frequently appended with stiffener components. **Figure 2** shows a 787's disassembled composite fuselage section which is composed of hat-stiffened composite panels that represent the design methodology of meeting the high stiffness while keeping the minimal weight requirements. This laminated composite stiffened panel is a critical component and extensively used structure in aircrafts, and can operate when subjected to harsh environments such as severe dynamic loading.

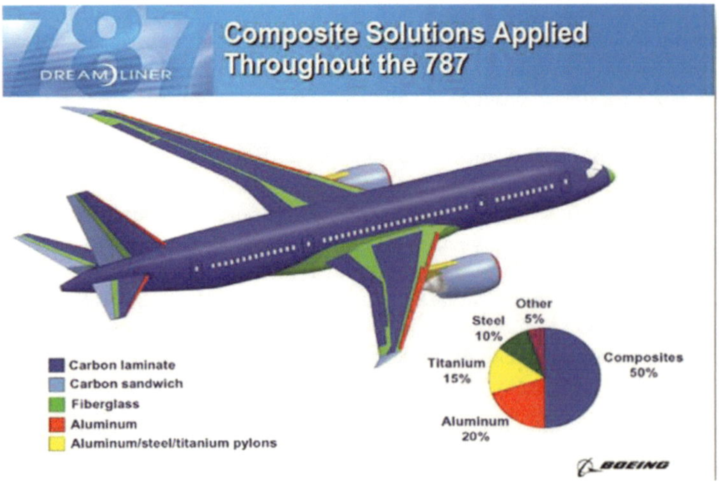

Figure 1. Boeing 787 aircraft contains 50% of composite materials.

Year	1982	1995	2006	2008
Model	Boeing 767	Boeing 777	Airbus 380	Boeing 787
Structures	Secondary	Primary/Secondary	Primary/Secondary	Primary/Secondary
Amount of CFRP/aircraft	1.5 tons	Approx. 10 tons	Approx. 35 tons	Approx. 35 tons
Amount of CF/aircraft	1 ton	Approx. 7 tons	Approx. 23 tons	Approx. 23 tons.

Table 1. Increase of carbon fiber composites for aircraft application.

Figure 2. Disassembled composite fuselage section of the Boeing 787.

Many work have been done on design and analysis of hat-stiffener structures. Recent advances in performing global and detailed analyses have made it possible to determine failure modes, strength, durability, and damage tolerance of composite structures with confidence. Bhar et al. [2] performed linearly elastic static and natural vibration analysis using an extended HSDT (higher-order shear deformation theory). Kim et al. [3] manufactured stiffened panels using co-curing, co-bonding and secondary bonding processes and evaluated them using 3D measurement and ultrasonic tester. Lauterbach et al. [4] built analysis tools including an approach for predicting interlaminar damage initiation and degradation models for capturing interlaminar damage growth as well as in plane damage mechanisms. Gangadhara et al. [5] analyzed stiffened panels using formulation based on the concept of equal displacements at the shell-stiffener interface. Kumar et al. studied the transient response of laminated stiffened plates using MSC/Patran and LS-DYNA3D [6] and Kristinsdottir et al. [7] presented an optimization formulation for the design of large panels when loads vary over the panel. Junhou et al. and Shenoi et al. [8, 9] examined the key aspects defining the performance characteristics of hat-stiffener joints in marine structures. Paul et al. [10] performed an integrated step-by-step design and analysis procedure for the hat-stiffened panels loaded in axial compression using the computer code BUSTCOP. Xiong and coworkers [11] has tested and analyzed the buckling and failure loads of hat-stiffened composite panels. Other research work have been focused on FEA modeling [12–22], manufacturing [23–29], evaluation of microstructures and damage evolution [30–33], and the enhancement of the mechanical properties [34–37] of composites at both materials- and structures-level.

Most commercial CAD/FEA software has included some form of parameterization of design variables. Basic research-level higher order structural elements are also developed. These tools allow quick, easy and accurate topology and geometry model creation with design constraints, implicit parameterization for easy model variation, integrated Finite Element generator, models and components storage in library for generation of knowledge database and reusability, shape and size optimization in a closed batch loop, on-the-fly definition of design variables and design space, and integration of specific applications like commercial optimization and design tools. In this work, we plan to utilize these aspects to create a Higher Order Abstract Structural Elements, later abbreviated as HOASE.

1.2. Objectives and structure

The goals and key feature of this work include analyzing the geometric parameter sensitivity of the hat stiffener, and developing and demonstrating a proof-of-concept theoretical model which is a parametric analytical solution that is theoretically equivalent to hat-stiffener stiffened panels in mechanical response. The analytical solution contains parametric information incorporating geometric, design allowables, and manufacturing information such as laminate stacking order. The constructions of these equivalent analytical models will be stored in a database from which they can be easily retrieved and parametrically modified.

Achieving the above requires specific technical objectives including:

1. Select composite ply materials and corresponding stochastic material properties for tracking them to parametric design allowables.
2. Explore the design space and using Finite Element Method (FEM) to analyze the parametrical sensitivity of the basic composite structural elements: hat stiffeners.
3. Develop an equivalent model using analytical solution and run case studies for various loading conditions to develop the empirical relationships between design parameters and allowables/performance. This takes into account the key geometric and material parameters and gives a higher and lower boundary of the relatively equivalent hat-stiffener stiffened panel.
4. Manufacture hat-stiffened composite panel and perform experimental investigation to compare its mechanical response with FEA models' prediction and the mechanical response bounds resulting from the analytical models. Finally, this work would provide the aviation industry with a parametric databases of hat stiffener design and analysis.

2. Optimal design procedure

2.1. Basic structural configuration and FEA design parameters

In this work, hat stiffeners and plates were selected as basic elements for parametric analysis and for constructing an analytical solution. The plate element is an orthotropic laminated element with material, number of plies, stacking sequence, width, length, and thickness parameters. The hat stiffener element is also parametrically defined in terms of several geometric parameters as shown in **Figure 3**.

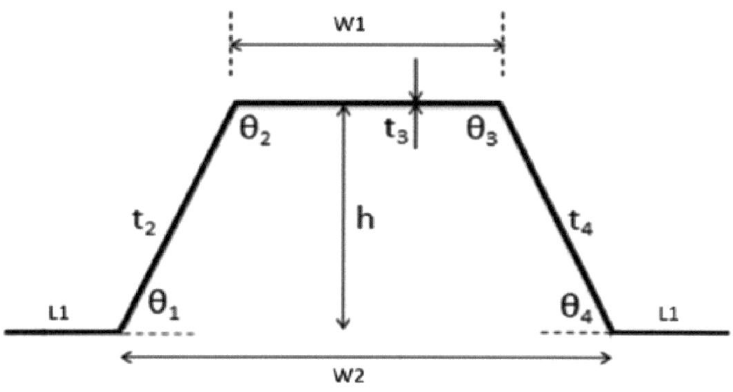

Figure 3. Hat stiffener basic element with geometric parameters.

FEM was utilized to produce sensitivity of structural behavior (deflection, stresses) to basic elements' parameters and for comparing final experimental results with modeling. Laminated plate, hat-stiffener, hat-stiffener bonded to base plate were modeled in MSC NASTRAN for this purpose. Laminated plate modeling in FEM is routine and therefore not discussed for the sake of brevity. The hat stiffener (with and without plate to which it is bonded) are modeled as follows. Height of the stiffener web (h), width of the stiffener cap (W_1), bottom width in between stiffener flanges (W_2), width of the stiffener flange (due to symmetric, the left and right width are the both L_1) are the geometric parameters considered in addition to thickness of a ply, ply orientation and the stacking sequence. The length of the hat stiffener is fixed at 508 mm (which is 20 inches).

Material properties are taken from Cytec information sheet CYCOM 5320 [12–37]. These unidirectional fiber tape tensile properties are:

E1 = 1.59E5 MPa;

E2 = 9.3E3 MPa;

Poisson's Ratio v = 0.336;

Shear modulus G12 = G13 = 5.6E3 MPa.

QUAD4 MSC Nastran element and PCOMP material properties input was used for analysis. A uniform pressure of 6.89E-2 MPa is applied on each of the two bottom flange surfaces for the hat-stiffener simulation. For the second set of simulations, same magnitude of pressure, 6.89E-2 MPa is applied on the plate to which hat-stiffener is bonded. Longitudinal edges are free to rotate but not translate ($T_x = T_y = T_z = 0$). The transverse direction edges are free. These longitudinal edge boundary conditions represent fixed edges rather than simply supported, because edge cross sections are constrained from rotation. Same boundary conditions for flat plates will represent simply supported conditions.

Longitudinal edges (the two edges of the skin plate only, not including hat stiffener web and top cap) are simply supported as $T_x = T_y = T_z = 0$ for hat stiffener bonded to the plate. The transverse edges of the plate are subjected to the boundary conditions $T_x = R_y = R_z = 0$, corresponding to all four edges simply supported. These boundary conditions are chosen to demonstrate extreme sensitivity of structural response to boundary conditions.

2.2. Parametric sensitivity analysis on hat-stiffener structures

To study the sensitivity of hat-stiffener's geometric parameters, hat-stiffener models are created first. The hat-stiffener element is modeled and analyzed using MSC NASTRAN to construct parametric design space. As presented in the last section, design parameters were defined for hat-stiffeners. The parametric range and increments we defined here covered most of the practical design exploration space and are summarized in **Table 2**.

These parametric variations represent 1680 models and design points. A smaller set of parameter combinations are analyzed to get the design trends. We explored maximum specific bending rigidity contribution of hat-stiffeners to membrane skin which is designed to take torsional shear. Representative 10 psi uniform pressure loading and simply supported boundary conditions on a 508 mm (20 inches) long hat cross section beam are analyzed. The cross-sectional area of hat-stiffeners varies with design parameters. A baseline configuration with minimum cross-sectional area is chosen to illustrate effect of parameters on bending. This configuration represents 12.7 mm (0.5 inch) bottom flange length, 25.4 mm (1 inch) bottom hat width, 12.7 mm (0.5 inch) top hat width, 12.7 mm (0.5 inch) hat height, 1.016 mm (0.04 inch) thickness and [0/90/45/−45]s stacking order.

Figure 4 shows the mid-point transverse deflection and maximum flexural stress at mid-point on the beam as a percent change from the baseline configuration. Stacking sequence and therefore corresponding laminate thickness is kept constant. The ratio of top and bottom hat widths is kept constant at 0.5 for all parametric variations. Three curve-sets show variation of deflection and flexural longitudinal stress with hat height, width and bottom flange length, respectively. As expected, it is evident that bottom flange length contribution is minimal to the

All dimensions are in **mm**

Parameter	Top Hat 1	Top Hat 2	Top Hat 3	Top Hat 4	Average	Std. Dev	CV (%)
Bottom Left Angle (θ_1)	50.2°	50.2°	51°	48.6°	50.0°	1.01°	2.01
Top Left Angle (θ_2)	130.3°	130.2°	129.4°	131.1°	130.3°	0.70°	0.53
Top Right Angle (θ_3)	130°	130.8°	131.0°	130.6°	130.6°	0.43°	0.33
Bottom Right Angle (θ_4)	49.5°	48.8°	48.6°	49.6°	49.1°	0.50°	1.02
Inside Length Top (l_t)	18.7	18.2	18.7	18.4	18.5	0.2	1.32
Inside Length Bottom (l_b)	59.4	61.1	61.7	62.5	61.2	1.3	2.15
Inside Height (h)	24.3	24.4	24.7	24.6	24.5	0.20	0.75
Thickness 1 (t_1)	1.11	1.13	1.14	1.09	1.12	0.02	1.98
Thickness 2 (t_2)	1.16	1.16	1.18	1.18	1.17	0.01	0.99
Thickness 3 (t_3)	1.20	1.24	1.19	1.21	1.21	0.02	1.79
Thickness 4 (t_4)	1.16	1.18	1.13	1.18	1.16	0.02	2.03
Thickness 5 (t_5)	1.10	1.11	1.09	1.15	1.11	0.03	2.36

Table 2. Hat-stiffener parametric design exploration space.

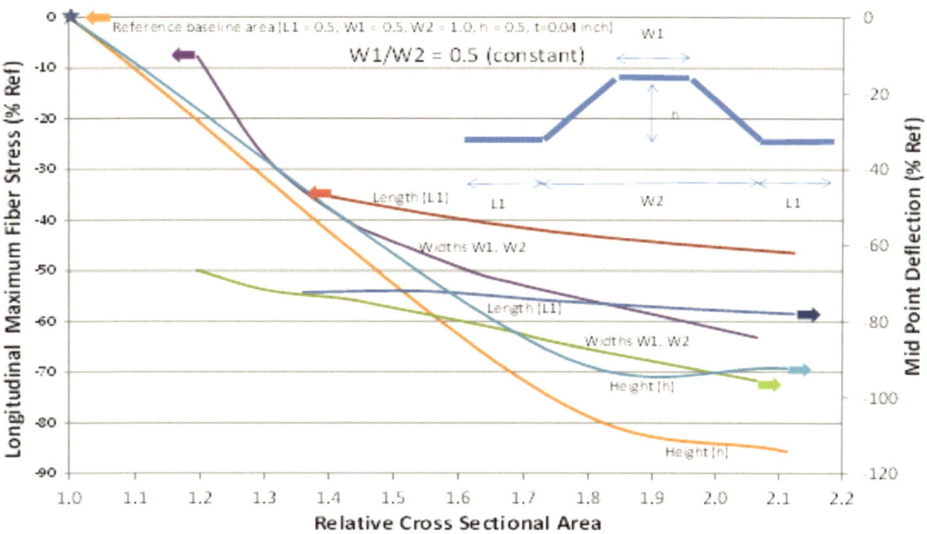

Figure 4. Hat stiffener basic element bending behavior.

flexural behavior of the stiffener. The maximum change in bending rigidity is achieved by changing hat height up to three times the top flange width.

2.3. Analytical solution of hat stiffener with base plate

A proof-of-concept analytical model consists of a rectangular plate stiffened by several hat stiffeners was established in MATLAB.

Figure 5 shows steps incorporated in constructing the analytical model. Composite ply properties, stacking sequence for hat and plate laminates, plate and hat stiffener geometric parameters, stiffener spacing, boundary conditions and loading are specified for the analytical model. Orthotropic plate properties are obtained by scaling, homogenizing and distributing stiffener properties over the space between the stiffeners.

Let θ be the angle between x-axis (stiffener longitudinal direction) and j is the ply fiber direction in sections plane, a is the equal distance between stiffeners. The bottom and top flanges as well as webs are defined as continuous plies of the orthotropic plate as follows:

For bottom flange:

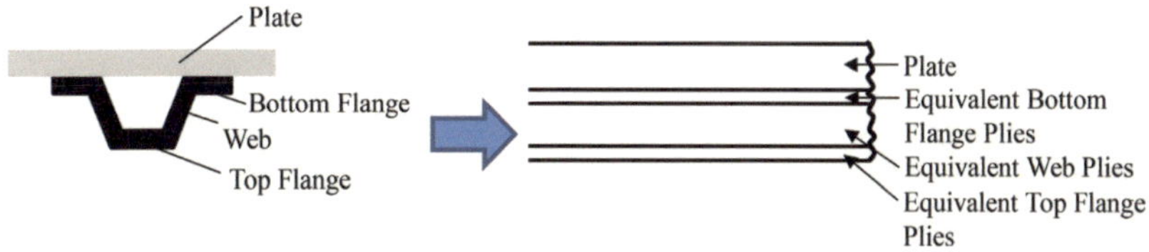

Figure 5. Equivalent orthotropic plate for hat-stiffener stiffened skin.

$$\overline{Q}^{j}_{11bf} = \frac{2L1}{a}\left[Q^{j}_{11}\cos 4\theta + Q^{j}_{22}\sin 4\theta + 2\left(Q^{j}_{12} + 2Q^{j}_{66}\right)\sin 2\theta\cos 2\theta\right] \quad (1)$$

For top flange:

$$\overline{Q}^{j}_{11tf} = \frac{w1}{a}\left[Q^{j}_{11}\cos 4\theta + Q^{j}_{22}\sin 4\theta + 2\left(Q^{j}_{12} + 2Q^{j}_{66}\right)\sin 2\theta\cos 2\theta\right] \quad (2)$$

For webs, define:

$$\cos x = \frac{h}{\left[\left(\frac{w_2 - w_1}{2}\right)^2 + h^2\right]^{1/2}} \quad (3)$$

And therefore, contribution from two web laminates is:

$$\overline{Q}^{j}_{11web} = \frac{2}{a\cos x}\sum_{j=1}^{n} t_j\left[Q^{j}_{11}\cos 4\theta + Q^{j}_{22}\sin 4\theta + 2\left(Q^{j}_{12} + 2Q^{j}_{66}\right)\sin 2\theta\cos 2\theta\right] \quad (4)$$

These equivalent $\overline{Q}^{j}_{11}$ contributions can be used in traditional ABD matrix construction. Similarly, other Qs, the equivalent reduced stiffness matrix components are also calculated, and their contributions are used in the traditional A, B and D matrix construction. The analytical solution of the equivalent panel was input into MATLAB for the center point deflection prediction. Future work will be focusing on analytically representing the homogenized panel equivalent to the stiffened panel with multiple hat-stiffeners on it. Also, it should be noted that for the analytical solution for large panel with a sparse distribution of multiple stiffeners, these relationships may not be valid but may still give the bounding values of the possible deflection of points on the plate.

2.4. FEA model for the demonstration part: Panel with multiple hat-stiffeners

To better understand and predict the mechanical behavior of the structure, a demonstration FEA model of one panel with multiple hat stiffeners bonded onto it was built in MSC Nastran (**Figure 6**). A few composite ply material properties were selected from the Cytec Cycom 5320 prepreg data sheet for creating the model and database.

For the geometric configuration of the model: this demonstrator model comprises a base panel of in-plane dimensions 304.8 mm (which is 12 inches) × 863.6 mm (which is 3 inches) with four hat stiffeners on it., each separated by approximately 85.725 mm (which is 3.37 inches). The bottom width of the hat stiffener is approximately 86.36 mm (which is 3.4 inches) with 60.96 mm (which is 2.4 inches) as the distance between the lower two corners of the hat stiffeners and 12.7 mm (which is 0.5 inch) overhang (i.e., flange) on the either side. The base panel has 8 plies of laminates with 5320 unidirectional prepreg properties and they are in a quasi-isotropic layup as follows: [90,−45,+45,0]S. Each of the four hats also consists of eight unidirectional fiber plies in the same quasi-isotropic layup.

http://dx.doi.org/10.5772/intechopen.79488

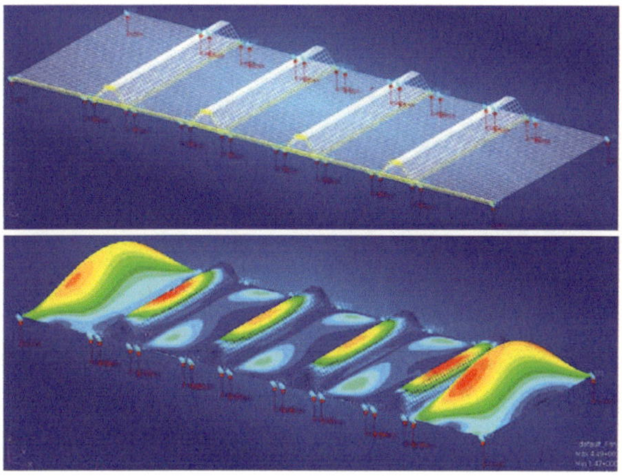

	Parameters
Number of Top Hats	4
Top Hat "W2"	57.15mm (2.25 inches)
Top Hat "W1" (always<W2)	15.875mm (0.625 inches)
Top Hat "h"	23.8125mm (0.9375 inches)
Top Hat "L1"	15.875mm (0.625 inches)
Width of base plate	863.6mm (34 inches)
Length in Y direction	For both Top Hat and Skin Plate: 304.8mm (12 inches)
t (Single Ply Thickness)	For both Top Hat and Base Plate: 0.127mm
Number of Ply	For both Top Hat and Skin Plate: 8 Plies
Stacking	For both Top Hat and Skin Plate: [90/+45/-45/0/0/-45/+45/90]

Figure 6. Midpoint deflection of the demonstration part FEA model.

Simulation of panel-level hat-stiffeners requires understanding of global and local effects of the parameters. One should consider local maximum deflection occurring in between the stiffeners on the panel, because that may become a dominant parameter for deformation constraints satisfaction.

2.5. Manufacturing of the demonstration part

To validate the modeling prediction of the center point deflection of the stiffened panel, a composite panel bonded with multiple hat-stiffeners was manufactured as a demonstrator part. During fabrication of the structural element and the final demonstration part, unidirectional Cytec Cycom 5320 prepreg material, out-of-autoclave curing, and secondary bonding technique were used.

The basic structural elements comprise of flat panels and hat cross section beams. The assembly of these basic structural elements forms the demonstrator part represented by a large panel stiffened by four equidistant hat beams, as shown in **Figure 7**. To accurately predict and compare with the FEA results, the demonstration part has identical set up with the MSC Nastran FEA model built and explained in the last section.

2.6. Testing and validation of the demonstration part

The demonstration part was tested under near-uniform 1 psi loading and the center point deflection was recorded so it can be compared with FEA results. Photographs of the testing setup are shown in **Figure 8**. The experimental testing of the demonstrator part involves simply supporting the edges and subjecting it to a uniform pressure loading condition by placing sandbags at the center. Experimentally measured panel displacements are then compared to predictions from both analytical constructs as well as FEA models.

The demonstration plate midpoint deflections are experimentally obtained for 150, 225, 300, 375 and 400 lb. load are 0.022, 0.032, 0.039, 0.045 and 0.047 in, respectively. The first increment

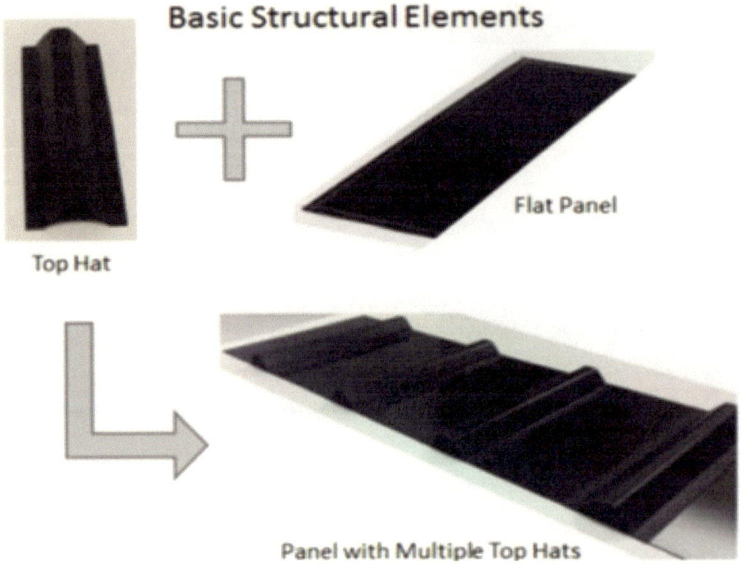

Figure 7. Manufacturing and assembly of basic structural elements into a demonstration part.

Figure 8. Simply supported hat-stiffened composite panel under near-uniform pressure loading.

(150 lb) was using lead balls filled bags providing close to uniform loading. The remaining increments were obtained using iron discs that did not provide as uniform loading as lead balls filled bags would have. As the results are shown in **Figure 9**, the midpoint deflection is 1.52 mm (0.06 in) for 1 psi uniform loading while FEA simulation gave 1.83 mm (0.07 in).

The analytical bounds for stiffened plates were also obtained. The midpoint deflection from the homogenized orthotropic plate gives the lower bound and simply supported idealized plate between the stiffeners gives upper bound. The lower bound provides better approximation for plates with closely spaced stiffeners. The real deformation starts to approach the upper bound as spacing between stiffeners increases. The lower bound for midpoint deflection under 1 psi is 0.133 mm (0.0052 in) and the upper bound is 2.85 mm (0.11 in).

The work performed establishes the basis for continuing future work to further develop a set of parametric models. The conceived process of designing advanced composite aircraft structural

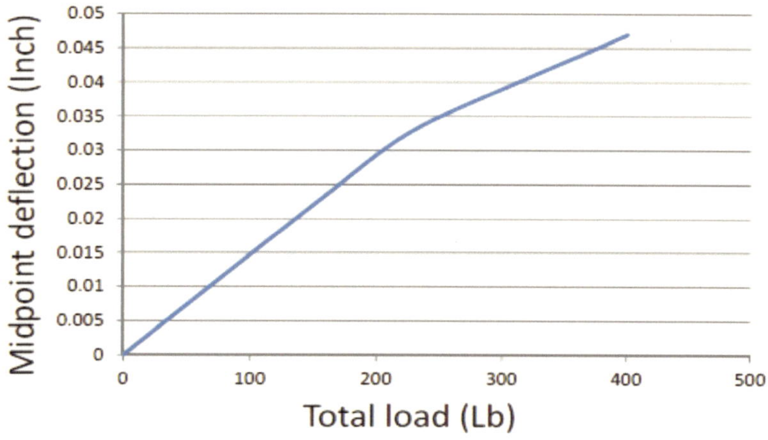

Figure 9. Midpoint deflection of the demonstration part.

components from these parametric modeling constructs will be matured, implemented and validated to demonstrate the benefits of starting the design with validated parametric design elements.

3. Conclusions

This work has illustrated the process of developing an analytical model and the design and analysis of the parametric composite hat-stiffened panels. The amount of the work involved in designing to this level of abstraction is a significant part of the design of an aircraft. This work is needlessly repeated by designers again and again and can be standardized to abbreviate the design process, and has successfully shown most of the processes involved in creating parametric models with a hat-stiffener stiffened composite laminated plate model development.

Most commercial CAD/FEA software includes some form of parameterization of design variables. Basic research level higher order structural elements have also been developed. These tools allow quick, easy and accurate topology and geometry model creation with design constraints; implicit parameterization for easy model variation; integrated Finite Element generator; models and components storage in library for generation of knowledge database and reusability; shape and size optimization in a closed batch loop; on-the-fly definition of design variables and design space; and integration of specific applications like commercial optimization and design tools. Our future work includes integrating these models in similar design tools, such as a combination of MSC Nastran, ABAQUS, MATLAB and C++ platform.

Acknowledgements

Authors would like to appreciate Brian Casey, Senior Engineer of MSC NASTRAN Development, for his important suggestions and the time he has spent on proof-reading the manuscript.

Author details

Bo Cheng Jin[1,2,3]*

*Address all correspondence to: bochengj@usc.edu

1 Department of Aerospace and Mechanical Engineering, University of Southern California, USA

2 Department of Chemical Engineering and Materials Science, University of Southern California, USA

3 MSC NASTRAN (NASA Structural Analysis) Development, USA

References

[1] Toray Industries. Market Research Report: Strategic Business Expansion of Carbon Fiber, Torayca. The Original; April 12, 2005

[2] Bhar A, Phoenix S, Satsangi S. Finite element analysis of laminated composite stiffened plates using FSDT and HSDT: A comparative perspective. Composite Structures. 2010;**92**:312-321

[3] Kim G-H, Choi J-H, Kweon J-H. Manufacture and performance evaluation of the composite hat-stiffened panel. Composite Structures. 2010;**92**:2276-2284

[4] Lauterbach S, Orifici AC, Wagner W, Balzani C, Abramovich H, Thomson R. Damage sensitivity of axially loaded stringer-stiffened curved CFRP panels. Composites Science and Technology. 2010;**70**:240-248

[5] Prusty BG. Linear static analysis of composite hat-stiffened laminated shells using finite elements. Finite Elements in Analysis and Design. 2003;**39**:1125-1138

[6] Kumar Y, Mukhopadhyay M. Transient response analysis of laminated stiffened plates. Composite Structures. 2002;**58**:97-107

[7] Kristinsdottir BP, Zabinsky ZB, Tuttle ME, Neogi S. Optimal design of large composite panels with varying loads. Composite Structures. 2001;**51**:93-102

[8] Junhou P, Shenoi RA. Examination of key aspects defining the performance characteristics of out-of-plane joints in FRP marine structures. Composites: Part A. 1996;**27**:89-103

[9] Shenoi RA, Hawkins GL. An investigation into the performance characteristics of top-hat stiffener to shell plating joints. Composite Structures. 1995;**30**:205-209

[10] Paul TK, Sinha PK. Design of hat-stiffened composite panels loaded in axial compression. Composite Structures. 1992;**21**:205-209

[11] Lee S, Xiong Y, Benak TJ, Heath J. Design, analysis, manufacture, and experimental performance of hat-stiffened composite panel. Science and Engineering of Composite Materials. 1997:37-50

[12] Jin B, Pelegri A. Three-dimensional numerical simulation of random fiber composites with high aspect ratio and high volume fraction. Journal of Engineering Materials and Technology. 2011;**133**:41014

[13] Wu H, Xu W, Shan D, Jin B. An extended GRN model for low stress triaxiality and application in spinning forming. Journal of Materials Processing Technology. 2019;**263**: 112-128

[14] Jain A, Jin B, Nutt S. Mean field homogenization methods for strand composites. Journal of Composites Part B. 2017;**124**:31-39

[15] Jin B, Li X, Jain A, Wu M, Mier R, Herraez M, et al. Prediction of stiffness of reused carbon fiber/epoxy composite oriented strand board using Finite Element Methods. Recycling Materials & Structures. Feature Article. Society for the Advancement of Material and Process Engineering (SAMPE) Journal. 2017;**53**:32-39

[16] Jin B, Li X, Jain A, Herraez M, Gonzalez C, LLorca J, et al. Development of a finite element model for reused carbon fiber epoxy composite oriented strand board. In: Proceedings of SAMPE 2016; Long Beach; 2016

[17] Jin B, Li X, Mier R, Pun A, Joshi S, Nutt S. Parametric modeling, higher order FEA and experimental investigation of hat-stiffened composite panels. Journal of Composite Structures. 2015;**128**:207-220

[18] Jin B, Joshi S, Pun A, Nutt S. Design sensitivity of hat-stiffened composite panels. In: Proceedings of American Society for Composites Conference 2014; San Diego; 2014

[19] Jin B, Liu W, Patel H, Nutt S. Application of MSC NASTRAN UDS in modeling and analysis of a hybrid composite reinforced conductor core. In: Proceedings of MSC NASTRAN Conference 2013; Los Angeles; 2013

[20] Jin B, Li X, Jain A, Wu M, Mier R, Herraez M, Gonzalez C, et al. Mechanical properties and finite element analysis of reused UD carbon fiber/epoxy OoA VBO composite oriented strand board. In: Proceedings of SAMPE 2016; Long Beach; 2016

[21] Jain A, Jin B, Li X, Nutt S. Stiffness predictions of random chip composites by combining finite element calculations with inclusion based models. In: Proceedings of SAMPE 2016; Long Beach; 2016

[22] Jain A, Jin B, Nutt S. Development of chip composites with improved mechanical performance attributes using hybrid multi-scale modelling methods. In: Proceedings of CAMX 2016; Anaheim; 2016

[23] Jin B. Recent Development of Reused Carbon Fiber Reinforced Composite Oriented Strand Boards. Chapter 8 of Book: Recent Developments in the Field of Carbon Fibers. Rijeka, Croatia: IntechOpen. DOI: 10.5772/intechopen.71346

[24] Jin B, Neidert K, Ellis M. Sustainable and efficient hydroforming of aerospace composite structures. Chapter of Book: Green Sustainable Composites. Rijeka, Croatia: IntechOpen. In Print

[25] Wu H, Fan G, Jin B, Geng L, Cui X, Huang M. Fabrication and mechanical properties of TiBw/Ti-Ti (Al) laminated composites. Journal of Materials and Design. 2016;**89**:697-702

[26] Chen Z, Huang T, Jin B, Hu J, Lu H, Nutt S. High yield synthesis of single-layer graphene microsheets with dimensional control. Journal of Carbon. 2014;**68**:167-174

[27] Huang T, Li T, Xin Y, Jin B, Chen Z, Su C, et al. Preparation and utility of a self-lubricating and anti-wear graphene oxide nano-polytetrafluoroethylene hybrid. Journal of RSC Advances. 2014;**4**:19814

[28] Jin B, Li X, Jain A, Gonzalez Carlos, LLorca J, Nutt S. Optimization of microstructure and mechanical properties of composite oriented strand board from reused prepreg. Journal of Composite Structures. 2017;**174**:389-398

[29] Wu M, Jin B, Nutt S. Processing and performance of recyclable composites for wind turbine blades. Advanced Manufacturing: Polymer & Composites Science. In Print

[30] Ma H, Xu W, Jin B, Shan D, Nutt S. Damage evaluation in tube spinnability test with ductile fracture criteria. International Journal of Mechanical Sciences. 2015;**100**:99-111

[31] Jin L, Niu Z, Jin B, Sun B, Gu B. Comparisons of static bending and fatigue damage between 3D angle-interlock and 3D orthogonal woven composites. Journal of Reinforced Plastics and Composites. 2012;**31**(14):935-945

[32] Ma Q, Wang K, Wang S, Liu H, Jin B, Jin L. Tensile damage mechanism of 3-D angle interlock woven composites using acoustic emission events monitoring. Autex Research Journal, ARJ-D-16-00031R1. 2017

[33] Jin B, Li X, Wu M, Jain A, Jormescu A, Gonzalez C, et al. Nondestructive testing and evaluation of conventional and reused carbon fiber epoxy composites using ultrasonic and stitched micro-CT. In: Proceedings of SAMPE 2016; Long Beach; 2016

[34] Wu H, Fan G, Jin B, Geng L, Cui X, Huang M, et al. Enhanced fracture toughness of TiBw/Ti3Al composites with a layered reinforcement distribution. Journal of Materials Science and Engineering: A. 2016;**670**:233-239

[35] Wu H, Jin B, Geng L, Fan G, Cui X, Huang M, et al. Ductile-phase toughening in TiBw/Ti-Ti3Al metallic-intermetallic laminate composites. Journal of Metallurgical and Materials Transactions A. 2015;**46**:3803-3807

[36] Huang T, Jin B, Li X, Li T, Nutt S. Experimental and finite element analysis study of load carrying capacity of modified polyimide. In: Proceedings of SAMPE 2013; Wichita; 2013

[37] Jin L, Jin B, Kar N, Nutt S, Sun B, Gu B. Tension-tension fatigue behavior of layer-to-layer 3D angle-interlock woven composites. Journal of Materials Chemistry and Physics. 2013;**140**:183-190

Chapter 5

The Guidelines of Material Design and Process Control on Hybrid Fiber Metal Laminate for Aircraft Structures

Sang Yoon Park and Won Jong Choi

Additional information is available at the end of the chapter

http://dx.doi.org/10.5772/intechopen.78217

Abstract

Fiber metal laminate (FML) is a hybrid material system that consists of thin metal sheets bonded into a laminate with intermediate thin fiber reinforced composite layers. The aerospace industry has recently increased their use of FMLs due to the considerable weight reduction and consequent benefits for critical load-carrying locations in commercial aircraft, such as upper fuselage skin panels. All FML materials and their processes should be qualified through enough tests and fabrication trials to demonstrate reproducible and reliable design criteria. In particular, proper surface treatment technologies are prerequisite for achieving long-term service capability through the adhesive bonding process. This chapter introduces a brief overview of design concept, material properties and process control methodologies to provide detailed background information with engineering practices and to help ensure stringent quality controls and substantiation of structure integrity. The guidelines and information found in this chapter are meant to be a documentation of current knowledge and an application of sound engineering principles to the FML part development for aerospace usage.

Keywords: fiber metal laminate (FML), mechanical behaviors, surface treatment, process control, long-term durability

1. Introduction: choice of materials in aircraft design

The current trends in commercial aircraft operations are showing an increasing demand for lower operational and maintenance costs. The maintenance costs, directly incurred by the airlines' operation, are an important measure of the economic benefits associated with reducing direct operating cost [1]. Practically, most aircraft structures are being designed for longer design lifetime with extension of inspection intervals. For this purpose, the fatigue and damage tolerance (F&DT) properties have been received considerable attention for further use of lightweight materials on next generation aircrafts [2]. Therefore, there is a strong need for the application of more durable and damage tolerant materials to minimize the total maintenance costs of commercial aircraft. Reducing the structural weight can lead to better fuel efficiency, reduced CO_2 emissions and lower maintenance costs. Nowadays, two competing materials, such as modern aluminum alloys and composites, have the potential to improve the cost effectiveness, but they still have limitations that restrict their widespread use, for example corrosion-fatigue resistance for aluminum alloys and blunt notch strength and impact resistance for carbon fiber reinforced plastics (CFRP) [3]. The basic characteristics of materials for aircraft structures are given in **Table 1**.

A chronological history of materials for aircraft structures is illustrated in **Figure 1**. New multi-layered hybrid materials, FMLs consist of thin metal sheets bonded into a laminate with intermediate thin fiber reinforced composite layers, and combines the benefits of both material classes [5]. Recently, the use of FMLs leads to subsequent benefits for primary aircraft structures, for example upper fuselage skin panels as shown in **Figure 2** [6–8]. This figure also presents typical load cases for dimensioning criteria in the design of fuselage structures. To date, the representative commercially available FML is glass reinforced aluminum laminate (GLARE), which combines thin aluminum sheets with unidirectional glass fiber reinforced epoxy layers [9, 10]. It has been produced for the upper fuselage skin panels of Airbus A380 (Toulouse, France) at GKN Aerospace's Fokker Technologies (Papendrecht, The Netherlands) in collaboration with AkzoNobel (Amsterdam, The Netherlands) and Alcoa (New York, US) [3, 4, 11]. The FMLs are also being considered for thin-walled structures for single aisle fuselage shells. In addition, their superior F&DT properties which are addressed as essential design principles in JAR/FAR 25.571 (Damage-tolerance and fatigue evaluation of structure) make them the ideal candidate for military aircrafts that such applications are not only subject to high fatigue stresses, but also high-velocity impact damages (e.g. battle damages) [12]. Concurrently, other types of commercially available FMLs are aramid aluminum laminate (ARALL) based on aramid fibers and carbon reinforced aluminum laminate (CARALL) based on carbon fibers, respectively [11].

Materials	Aluminum alloys	Composites (CFRP)	FML

Strength	• Broad experience • Repairability • Static behaviors • Improvement potential	• Fatigue behaviors • Low density (1.54 g/cm^3) • No corrosion • Best suited for smart structures	• Improved fatigue • Better tailoring • Higher fire resistance • Less corrosion
Weakness	• High density (2.78 g/cm^3) • Fatigue behaviors • Corrosion behaviors • High costs of new alloys	• Poor impact behaviors • No plasticity • Reparability • Recycling	• Lower stiffness • Higher density (2.52 g/cm^3) • Less industrialized process (compared to CFRP)

Table 1. Strength and weakness of materials for aircraft structures [4].

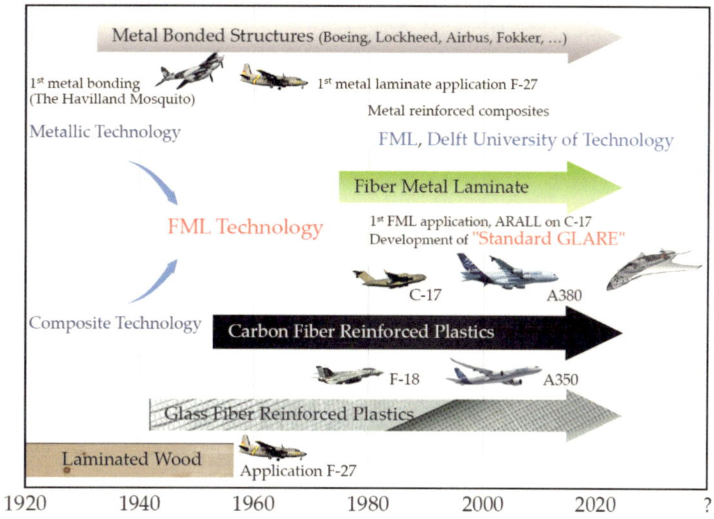

Figure 1. Chronological history of materials for aircraft structures (reproduced from Fontain [3]).

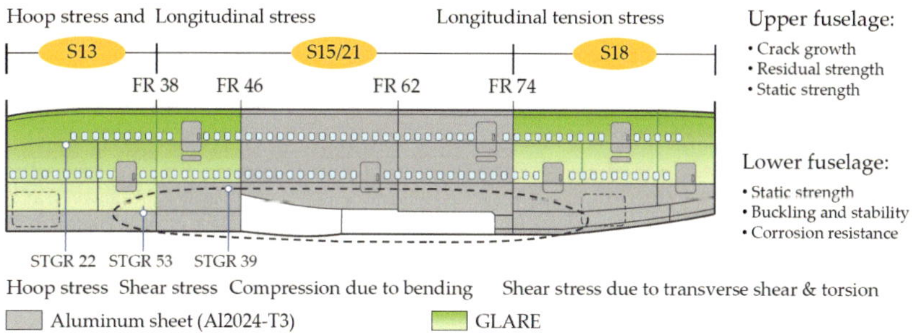

Figure 2. GLARE application on Airbus A380 fuselage section-13/18: Total GLARE area is 469 m^2, 27 panels (reproduced from Beumler [4]) and typical load cases on GLARE sections (reproduced from Assler and Telgkamp [13]).

The first generation FML, the ARALL, was introduced at 1978 in Faculty of Aerospace Engineering at TU Delft (Delft University of Technology, The Netherlands) [14]. The ARALL consists of alternating thin aluminum alloy layers (0.2–0.4 mm) and uniaxial or biaxial aramid fibers. The GLARE which is the second generation of FML presents the excellent fatigue resistance with high blunt notch strength than either 2024-T3 or ARALL. This new hybrid material also offers the actual weight reduction when it is applied to the fuselage skin panels [15, 16]. Finally, a much stiffer FML which is made by carbon fiber instead of aramid and glass fibers, the CARALL, had been also investigated in TU Delft [17]. The use of high modulus of carbon fiber (in typical, ranging from 230 to 294 GPa) exhibits more efficient crack bridging at the preliminary stage of fatigue crack propagation within composite layers [18]. However, the residual strength of notched CARLL is significantly lower than the monolithic aluminum alloys due to the limited failure strain of carbon fiber (in typical, 2.0%) [11]. Furthermore, it is more susceptible to galvanic corrosion when aluminum alloys are electrically connected to carbon fiber reinforced composites [19–21].

2. Material property behaviors of GLAREs

2.1. Mechanical behaviors of GLAREs for aircraft structures

GLAREs boast a large number of favorable characteristics, such as low density, static strength, better F&DT properties, high impact and flame resistances, as shown in **Figure 3** [22–24]. More descriptions on advantages of GLAREs are provided as follows:

- *Lightweight*: High static strength of GLARE2 in 0° fiber direction contributes to weight saving over the aluminum alloys by roughly 6% in the design based on bending stiffness, and by 17% in the design based on yield strength, respectively [25]. For example, the use of GLAREs on A380 fuselage shells achieves a weight saving of up to 794 kg (−10%) compared with the aluminum alloys [26]. The typical density of standard GLAREs is the range from 2.38 to 2.52 g/cm^3.
- *High strength*: It is apparent that the GLAREs reinforced with unidirectional glass fiber have anisotropic properties. This glass fiber contributes to increase in static strength and elastic modulus in the longitudinal direction along which the glass fiber is oriented. On the other hands, the aluminum sheets control overall mechanical properties of GLAREs in the transverse direction. As a result, the unidirectional GLAREs (e.g. GLARE1 and GLARE2) exhibit the high ultimate tensile strength compared with the aluminum alloys in the longitudinal direction, and it contributes to weight reduction in the case of tension-dominated structural components. In contrast, the transverse strength is somewhat lower than those of aluminum alloys. To overcome this limitation, the cross-ply GLAREs (e.g. GLARE3 and GLARE5) and angle-ply GLARE (e.g. GLARE6) have been introduced to provide the balanced mechanical properties in both directions [26].
- *High fatigue resistance*: The superior fatigue resistance is a result of fiber bridging mechanism whereby the intact fiber layers provide an alternative load path around the cracked metal layers, eventually reducing local stress in front of crack tip [27]. Vogelesang et al.

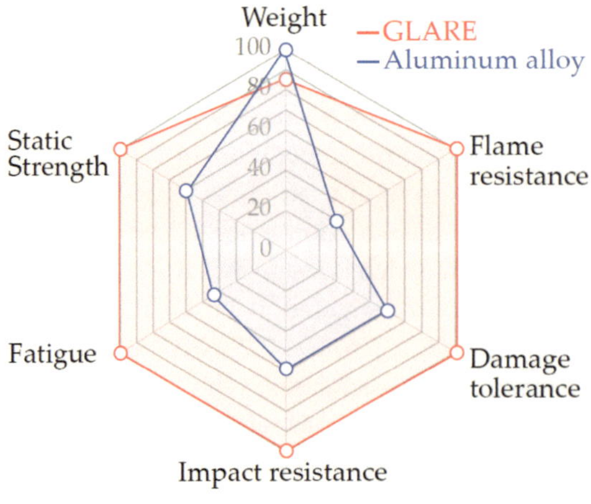

Figure 3. GLARE vs. aluminum alloy comparison ratio.

[28] reported that GLARE3–3/2 exhibit almost constant slow crack-growth when it is subjected to constant-amplitude fatigue loading as shown in **Figure 4**. Such low fatigue crack growth rate can lead to the minimal scheduled inspection downtime of aircraft.

- *Blunt notch strength*: The notched residual strength is also an important design parameter since the geometrical notches (e.g. cutouts to serve as doors and windows) are inevitable in the design of fuselage shells. Although the GLARE presents a relatively high notch sensitivity compared with ductile aluminum alloys, and the use of high ultimate strength S2-unidirectional glass fiber (in typical, 4890 MPa) makes it superior to ARALL in notch strength [26]. Hagenbeek et al. [29] proposed a numerical simulation model for predicting blunt notch strength by considering the metal volume fraction based on Norris failure criterion, and they reported that such approach has been shown to be effective for use in predicting multi-axial blunt notch strengths (i.e. biaxial and shear components) of GLARE.
- *High impact resistance*: Impact resistance is actually a significant advantage of GLARE, especially when compared to either aluminum alloys or CFRP [30]; **Figure 5** compares the respective impact energy absorbing capacities based on the through-the-thickness cracking (i.e. puncture energy). Obviously, GLARE3–3/2 shows higher impact resistance to cracking than aluminum alloy. This result may be attributed due to the localized fiber fracture and the extensive shear failure in the outer aluminum sheets [31, 32]. In addition, a high strain rate strengthening phenomenon which occurs in the glass fibers, combined with their relatively high failure strain contribute to increase in the impact resistance of GLARE, rather than other FMLs, such as ARALL and CARALL [33].
- *Burn-through capabilities*: To meet the airworthiness standard of a max. 90 seconds evacuation time (JAR/FAR 25.803: Emergency evacuation), a structural integrity of fuselage is of major importance in order to prolong a safe environment of the passengers in the event of a post-crash fire scenario. Apparently, the GLARE shows high thermal insulation

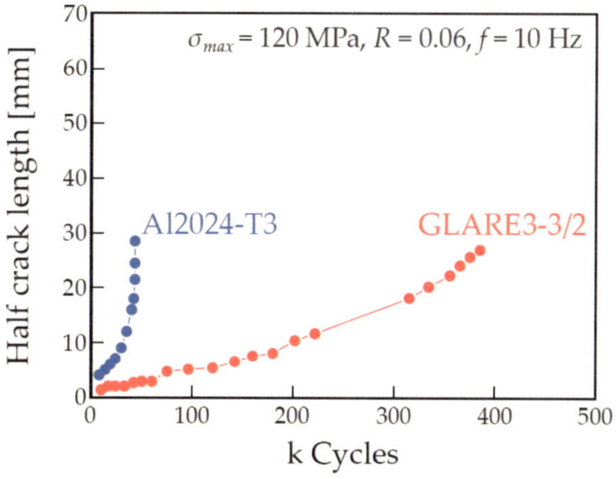

Figure 4. Fatigue crack growth [14].

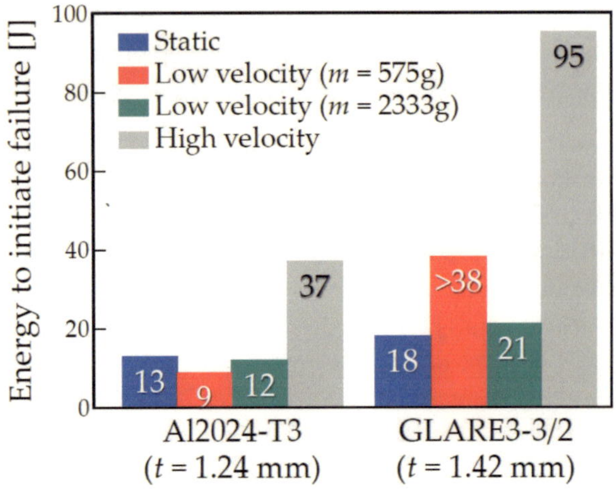

Figure 5. Comparison of low-velocity impact performance [15].

performance, and subsequently contributes to enhancing structural integrity in fuselage shells as shown in **Figure 6**. Owing to high melting temperature of S2 glass fiber (in typical, 1466°C), only the outer aluminum sheet starts to melt and separates from the other layers. As a result, the unexposed side of GLARE panel would remain relatively intact where the unexposed side temperature was just below 400°C.

- *Long-term hygrothermal behaviors*: In general, the significant changes in moisture absorption are not observed by GLAREs, which confirmed well to the shielding effect of the outer aluminum sheet in this material [35, 36]. However, in the case of thermal cycling exposure, the decrease rate of GLAREs is 1–7% higher than those of glass fiber-reinforced composites. This reduction is attributed to the large coefficient of thermal expansion (CTE) difference between their constituent materials [35].

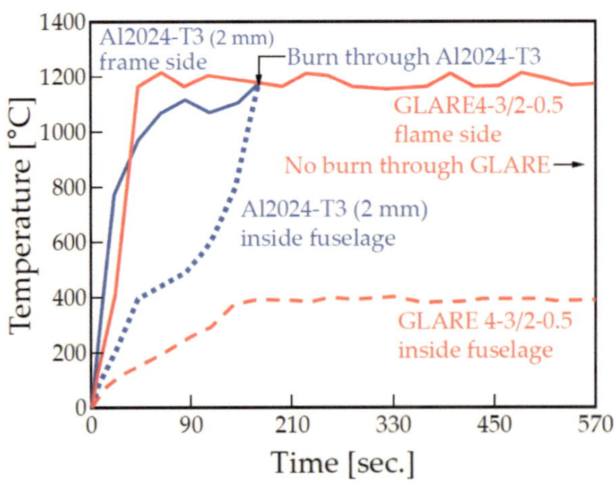

Figure 6. GLARE fire resistance comparing to aluminum alloy [34].

2.2. GLARE grades

Another beneficial feature of GLARE is that the number and orientation of composite layers can be selected to best suit different applications, and such material features make it attractive for structural applications [37]. For the certification of GLAREs for aircraft structures, several lay-up patterns are already defined as a standard grade: the schematic view of GLARE 3/2 is shown in **Figure 7**. This approach is useful to define the specific lay-up pattern used in the structural design [38]. Nowadays, the standard GLAREs are being produced in six different grades as listed in **Table 2** [39]. All grades are classified according to the type of lay-up pattern where the composite layers consists of unidirectional S2 glass fiber (AGY Holding Corp., USA)

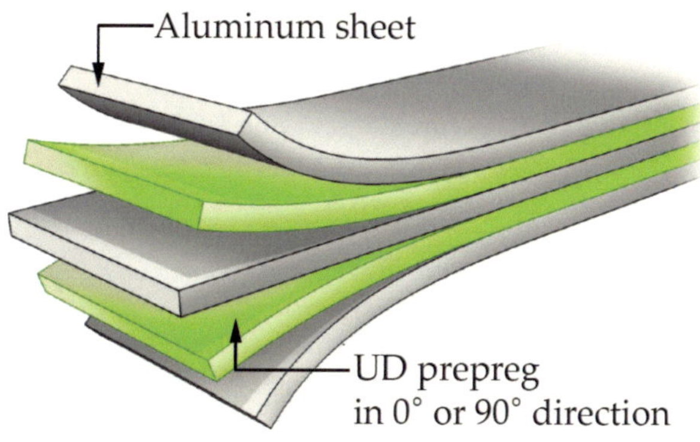

Figure 7. Schematic view of GLARE 3/2.

Grade	Metal layers		Prepreg layers		Typical density (g/cm³)	Characteristics
	Grade	Thickness (mm)	Orientation	Thickness (mm)		
GLARE1	7475-T761	0.3–0.4	0/0	0.25	2.52	• Unidirectional loaded parts with rolling direction aluminum sheet in loading direction (stiffeners)
GLARE2	2024-T3	0.2–0.5	0/0	0.25		
	2024-T3	0.2–0.5	90/90	0.25		
GLARE3	2024-T3	0.2–0.5	0/90	0.25	2.52	• Bi-axially loaded parts with 1:1 of principle stresses (fuselage skins, bulkheads)
GLARE4	2024-T3	0.2–0.5	0/90/0	0.375	2.52	• Bi-axially loaded parts with 2:1 of principle stresses with aluminum sheet in main or perpendicular loading direction (fuselage skins)
	2024-T3	0.2–0.5	90/0/90	0.375		
GLARE5	2024-T3	0.2–0.5	0/90/90/0	0.5	2.38	• Impact critical areas (floors & cargo liners)
GLARE6	2024-T3	0.2–0.5	−45/+45	0.5	2.52	• Shear, off-axis properties

(a) The number of orientations is equal to the number of unidirectional prepreg ply in each composite layer. The thickness in mm corresponds to the total thickness of composite layers in between two aluminum layers.
(b) The rolling direction (axial) is defined as 0°, and the transverse rolling direction is defined as 90°.

Table 2. Classification of GLARE for aircraft structures [5, 12, 39].

and FM®94 modified epoxy (Cytec-Solvay Group, USA). Nominal fiber volume fraction and ply thickness of prepreg are 59% and 0.125 mm, respectively [39].

A special coding convention is used to describe the different GLARE grades and specify their lay-up patterns. Symbolically, a general configuration is represented as follows [5]:

$$GLARE\ N_G = N_{al}/N_{gl} - t_{al} \tag{1}$$

where; N_G is the number indicating GLARE grade, N_{al} is the number of aluminum layers, N_{gl} is the number of composite layers ($N_{gl} = N_{al-1}$) and t_{al} is the thickness of each separate aluminum sheet (in typical, 0.25–0.5 mm). Each composite layer in turn consists of a certain number of unidirectional prepreg plies in 0°/90°/±45° directions. For example, each composite layer in GLARE4 consists of two unidirectional prepreg plies in oriented at 0 and 90°with respect to the rolling direction of aluminum sheets. Thereafter, GLARE4B-3/2 comprises three cross-plies within a composite layer, for example two layers in 90° and one layer in 0° direction. The fraction of unidirectional fibers in the rolling direction is twice much than that in the perpendicular direction.

2.3. Design philosophy for GLARE structures

An introduction of new materials for aircraft structures took place in evolution steps which suggests a realistic application of the innovative design philosophy, eventually leading to optimization of design concept. The innovative design concept of GLAREs on A380 fuselage shells is shown in **Figure 8** [40]. The structural efficiencies, such as damage tolerance and residual notch strength are much better served by incorporating the local variations in skin panel thickness with adhesively bonded joints. In early stage of technology development, GLARE structures were produced only as a flat panel. The innovations in structural design have been developed to overcome the joining problem and is termed the splice joint. The first splice concept consisted of butted aluminum sheets with the composite layer bridging the splices (e.g. butt joint). However, this concept is not recommended for structural applications because of premature failure in the butts. To overcome this limitation, several designs of splice concepts where two aluminum sheets are positioned with a slight over-lap forming a single metal sheet layer are introduced, as shown in **Table 3** [41].

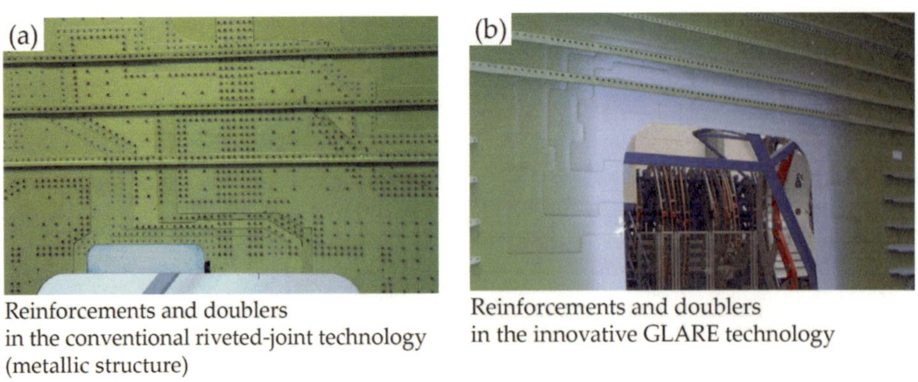

Figure 8. Construction and production possibility with the optimized GLARE panel (reproduced from Wischmann [40]).

http://dx.doi.org/10.5772/intechopen.78217

Splice in skin panel or doublers 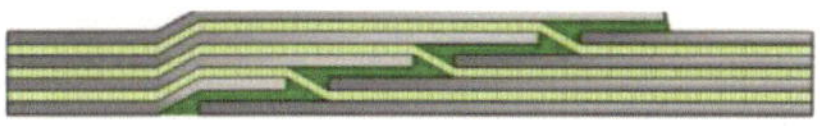 • Lamination of aluminum sheet width • Internal stress level in double curved panels	Additional glass fiber layers 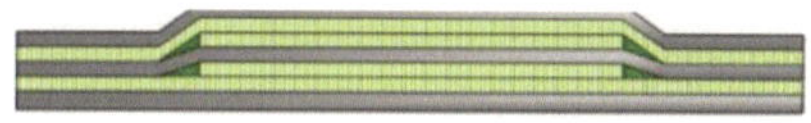 • Embedded at frame locations
Inter-laminar doublers 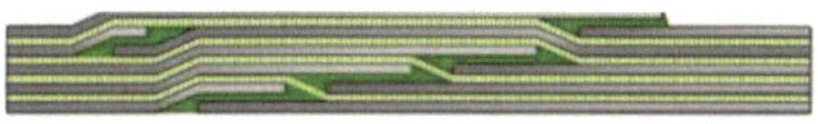 • Spliced, go through depending on length/orientation	Fiber oriented and lay-up 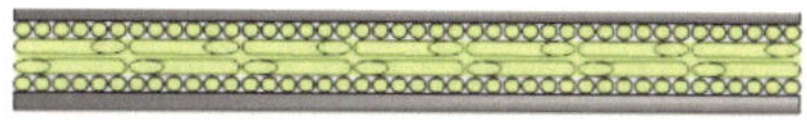 • Adjust properties to loading condition
Transition of GLARE type 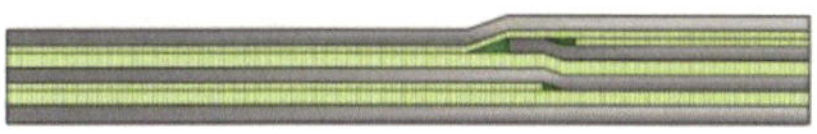 • e.g. GLARE4 to GLARE3	Additional layers: 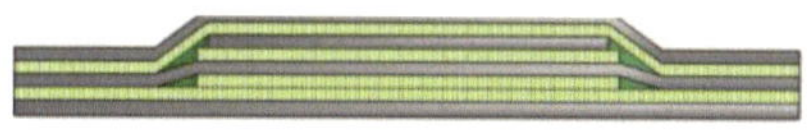 • Aluminum sheet locally at frame station • Glass layers locally between two aluminum sheets

Table 3. GLARE design features–"*giant tool box*" (reproduced from Pleitner [41]).

For example, alternating layers (i.e. aluminum sheets and unidirectional S2 glass/epoxy prepreg plies) are laid up over the curing tool, in which forms single, or double curvature shape [5, 42]. The splices are then staggered with respect to each other, while the adhesive layers are continuous. This interlaminar doubler solution can offer the local thickness variations in the skin panel [43]. Furthermore, this design concept can allow tailor-made skin panels of any size, not limited by the standard width of aluminum sheet rolls (in typical, 1.5 m). Now, the practical limitation of panel sized is only defined by autoclave size. The thickness variations in the skin panel are generally utilized for compliance with the fail-safe design requirements and the cost-effective part production for integrating the fuselage structures between skin panels, longitudinal stringers and circumstance frames. However, the difficulties in the production of splice GLARE panels in two bonding cycles demand for a feasible production solution, which allows for completing a splice joint including doublers through co-curing process. For this purpose, a SFT (Self-Forming Technique) can provide a smart solution to produce the required doublers without an additional cure cycle for bonding the doublers over the base GLARE panel. Such an inter-laminar panel highlights the advantages of using a SFT process as follows: (1) no dimensional tolerance issue for overlap in double curvature panel, (2) the evacuation of entrapped air or volatiles in composite layers through splice (adhesive squeeze-out). It therefore allows for the increased fuselage panels width with reducing the additional joints, structural weight and production cost [5].

2.4. Metal volume fraction (MVF)

For the standard GLARE grades qualified, their in-plane static properties can be defined by simple prediction based on MVF, which can reduce the additional experimental testing for material qualification. A terminology, MVF, reflects the relative contribution of aluminums properties to the properties of GLARE [5]. As a result, the MVF approach is useful for the prediction of static strength properties for GLARE as found in the literatures [5, 44, 45]. The MVF value can be calculated as follows:

$$MVF = \sum_{p_{metal}}^{t} t_{al}/t_{laminate} \tag{2}$$

where; t_{al} is the thickness of each separate aluminums sheet, $t_{laminate}$ is the total thickness of GLARE panel and p_{metal} is the number of aluminums sheets [5]. The typical MVF values of the standard GLARE grades are valid in a range between 0.55 and 0.70. The material property of GLARE having any MVF can be calculated by using a linear relation which follows the "*rule of mixtures*" available in anisotropic mechanics by using the Eq. (3).

$$E_{GLARE} = E_M \cdot MVF + E_G(1 - MVF) \tag{3}$$

where; E_{GLARE} is the elasticity of GLARE and E_M and E_G are the elasticity of aluminum sheet and composite layers, respectively. The load transfer ratio for composite layers (P_G/P_{GLARE}) in GLARE according to MVF can be defined as follows:

$$\frac{P_G}{P_{GLARE}} = \frac{E_G/E_M}{E_G/E_M + MVF/(1 - MVF)} \tag{4}$$

The load transfer ratio for composite layers in GLARE according to MVF can be predicted as shown in **Figure 9**. It is worth noting that the load transfer ratio of composite layer in GLARE exponentially decreases with the fraction of aluminum sheets. As the fraction of aluminum sheets in GLARE decreases, more shear load can be dissipated through the aluminum sheet-composite interface [6].

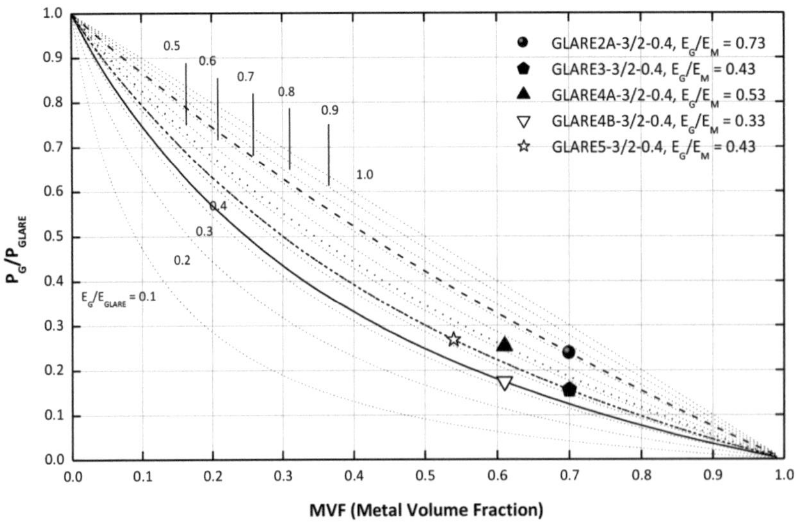

Figure 9. Plot of load transfer ratio for glass/epoxy layers in GLARE according to MVF for various modulus ratios of E_G/E_M: The corresponding GLARE grades of • GLARE2A 3/2–0.4 (0.703), ⬟ GLARE3 3/2–0.4 (0.703), ▲ GLARE4A 3/2–0.4 (0.612), ▽ GLARE4B 3/2–0.4 (0.612) and ☆ GLARE5 3/2–0.4 (0.542).

3. Process control methodologies for producing GLARE structures

3.1. Part production and quality controls

Basically, the production process for making GLARE structures is similar to the traditional production of metallic bonded structures and composite laminates. Before the parts are released, the part's quality should be assured through a reliable quality control (QC) method. For this purpose, the stringent QCs procedures shall be developed and applied to the part production of GLAREs. At this time, the QCs system includes all procedures that ensure the raw material quality, in-process control monitoring and verification of fitness for part acceptance. At each production stage, the key process parameters should be also standardized with the specified production tolerances as the follows [4, 5, 41]:

- *QC of raw materials*. GLARE manufacturer starts with the preparation with rolls of thin aluminum bare sheet (in typical, 0.3–0.4 mm). A custom-built machine decoils the thin aluminum sheet from rolls, and flattens the sheet and cuts it to lengths of up to 11 m for large skin panels. Next, the cut sheets are milled in accordance with the engineering drawings. At this time, all the aluminum sheets and unidirectional prepreg plies should be controlled by raw materials inspection specifications, and some specific properties should be controlled: (1) rolling direction, straightness, waviness and surface roughness for aluminum sheets; (2) fiber direction, prepreg bridging, or wrinkles and shelf-life requirements (e.g. storage life and mechanical life) for prepreg plies. This QC activity is basically the same as the traditional production of sheet metal forming, or composites. All prepreg shall be cut over a clean, non-contaminated surface with clean, sharp knives, or digital cutting machine to minimize distortion and splitting. The pre-cut materials (i.e. kit) should be stored in flat or stress-free condition to prevent folding or further damage. Unless otherwise specified by the engineering drawings, all the prepreg size should have a suitable trim at required locations to keep irregular edges out of the final part dimension.

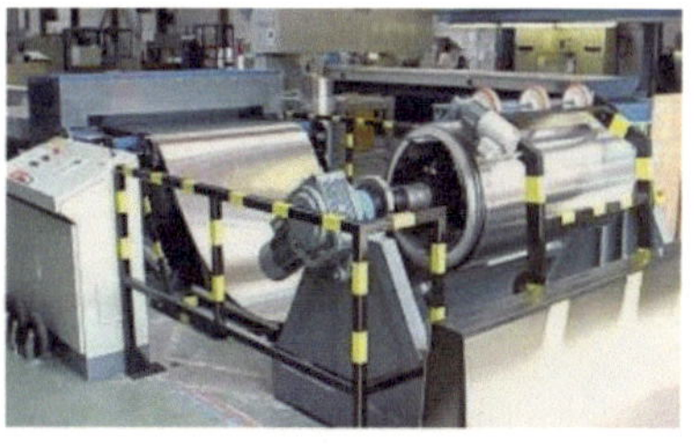

[Source: CompositesWorld]

- *QC of surface treatment*. Surface of aluminum bare sheet should be pre-treated to obtain a proper adhesion strength and durability with the prepreg resin. For this purpose, the milled sheets are transferred via a handling system to the chemical treatment line. The standard surface treatment process consists of solvent degreasing, CAE (Chromic-Sulfuric Acid Etching), CAA (Chromic Acid Anodizing), and followed by organic bond primer. All key process parameters should be checked for each batch of aluminum sheets according to the corresponding Airbus's own specifications, for example deoxidizing/anodic bath temperature, solution chemistry, rinse water purity and so forth. The specific surface treatment procedures of aluminum alloys are going to be explained in detail Section 3.2. Finally, the primed, cut sheets are re-rolled, and covered in a protective black plastic (or paper) bag for storage until needed in fabrication.

[Source: Fokker]

- *Control of lay-up process*. Alternating layers of aluminum bare sheets and prepreg plies are positioned in the right stacking order in accordance with GLARE grade. All the lay-up works should be conducted in a sufficiently clean environment, and the working environment such as temperature and humidity should be also kept below well-defined levels. All cut prepreg plies should be sequentially prepared and collated on the curing tool in the location and orientation as per the engineering drawings, or shop process instruction. An optical LPS (Laser Projection System, Virtek Vision International Inc., Waterloo, ON, Canada) may be capable of attaining the required dimension tolerance.

[Source: Fokker]

- *Control of autoclave process*. The laid-up parts are vacuum bagged, and then placed into the autoclave to be united by heat and pressure. Autoclave facility shall have instrumentation which autographically records time, temperature, pressure and vacuum where applicable. All gauges shall be controlled and periodically calibrated and certified in accordance with the procedures approved by the QC department. During an autoclave cure cycle, a high compaction pressure (in typical, 11 bar) is normally applied to the GLARE lamination stack at an elevated curing temperature (in typical, 125°C) for 3.5 hours. The representative manufacturing-induced defects, such as voids, porosities, should be accurately controlled to prevent internal defects. In addition to the QC activities in the part production of GLARE, there is also required to perform a *"final check"* prior to the part release. Non-destructive inspection (e.g. ultrasonic C-scanner) and some mechanical tests are generally accomplished in the final step of QC.

[Source: CompositesWorld]

- *Post processing*. The manufacturing and assembly of GLARE structures typically require machining operations, such as milling and drilling. For examples, the GKN Aerospace's Fokker business has been produced a large-sized GLARE panel of 4.5 × 11.5 m by using a 5-axis milling machine on a movable bed. However, a technique for machining of this multi-layered structure has presented more challenges in the aerospace industry than aluminum alloys or composites due to the coupled interaction between composite- and metal-phase cutting. The machining operations should be accomplished to meet the acceptance limit for the discrepancies as per the engineering drawing, or process specification.

[Source: CompositesWorld]

3.2. Surface treatment of aluminum alloys for producing GLARE structures

A strong bonding interface is one of the key factors for improving the durability of GLAREs. It is apparent that the surface treatment technique can improve the surface energy and wettability of metallic substrates, and is an effective method for enhancing the bonding strength between a metal substrate and a fiber reinforced polymer composite [46]. In addition, the surface treatment can remove the undesirable surface oxides or contaminants on the metallic substrate, and ameliorate the surface composition and microstructure of the metallic substrate [6, 47]. This allows the fiber bridging mechanism and mechanical properties of GLAREs to be improved, and moreover, the crack propagation rate at the aluminum-composite interface can be effectively reduced [48, 49]. Previous research works reported that the surface treatment should be carefully taken into consideration when improving interlaminar shear strength at the aluminum-composite interface [6, 50], environmental durability and low-velocity impact resistance of GLARE. Therefore, the proper production steps should be clearly defined before any production process is implemented. Note that this section is described based on our previous surface treatment studies of aluminum alloys for aircraft structures [46].

All the anodizing process are complex multi-stage operations incorporating degreasing and deoxidizing stages, as described in the preceding sections, plus appropriate rinses. Anodizing oxidation in solution of CAA or PAA is the preferred stabilizing treatment for the structural adhesive bonding of aluminum alloys in critical applications such as aircraft structures [51, 52]. However, they typically rely on such hazardous materials as strong acid and hexavalent chromium. The use of chromate is prohibited, or progressively banned in most industries due to its carcinogen activity. For this purpose, non-chromate anodizing such as boric-sulfuric acid anodizing (BSAA), or phosphoric-sulfuric anodizing (PSA), have been developed since the mid-1990s [51, 53], but neither of them have been fully validated for aircraft applications. Typical anodizing processes and their process parameters are listed in **Table 4**.

The classical porous oxide structure which are produced by anodizing process is likely to be related to capillary forces of primer trying to penetrate into the oxide pores, which in turn increase in mechanical interlocking between anodic oxide and primer [46]. The porous oxide structures can be controllable in accordance with the anodizing processes, as listed in **Table 5**. This table clearly represents the effects of anodizing processes on the oxide structures in terms of oxide thickness, pore diameter and cell wall thickness. The CAA process was found to give a relatively thick and softer oxide structure than those formed by the other processes [52]. This was established as an effective pretreatment for adhesive bonding with superior durability performance in service [51, 52, 54]. The European aerospace industry is still using this method [51, 52]. However, notwithstanding the remarkable durability data in corrosive environments, the use of chromate treatment process is being restricted due to recent environmental policy.

The PAA process is basically used for the structural adhesive bonding of aluminum and its alloys. The standard process (Boeing's BAC 5555 or ASTM D 3933) has proven to produce the most durable and reactive surface for structural adhesive bonding [57]. The PAA substrates are normally submitted to a Forest Product Laboratory (FPL) etch prior to anodizing, although the non-chromate acid etch (P2) is sometimes used instead. The PAA-treated anodic oxide is highly porous with open cell diameter of approximately 32 nm in height on top of a much

http://dx.doi.org/10.5772/intechopen.78217

Treatments[1]	CAA [51]	PAA [54]	BSAA [51]	PSA [53]
Electrolyte (wt%)	2.5–3.0 (CrO_3)	10 (H_3PO_4)	5.0–10.0 (H_3BO_3)/ 30.0–50.0 (H_2SO_4)	10.0 (H_3PO_4)/ 10.0 (H_2SO_4)
Voltage (V)	40.0 ± 1.0	10.0	15.0 ± 1.0	18 ± 2.0
Time (min)	35–45	20	18–22	15
Temperature (°C)	40.0 ± 2.0	25.0	26.7 ± 2.2	27.0 ± 2.0
Contamination controls	• Control Cl-[2] & sulfate impurity • Incorporation of $BaCO_3$ powder[3] to remove impurity	• Control Cl- & F[4] • Filtering required to remove fungus	• Prone to biological contamination[5] • Use of sodium benzoate or benzoic acid to prevent fungus growth	• Control Cl- & F • The installation of preventive devices for fungus growth (e.g.filters and UV lamps)
Racks	Al, Ti, Al with Ti-tips	Equivalent to CAA	Equivalent to CAA	Equivalent to CAA
QC issues	• Appearances • Solution chemistry • Water purity • Air cleanliness • Voltages • Bath temperature	• Appearances • Solution chemistry • Water purity • Air cleanliness • Voltages • Coating weight	• Appearances • Solution chemistry • Water purity • Air cleanliness • Voltages	• Appearances • Solution chemistry • Water purity • Air cleanliness • Voltages

[1]The proprietary materials and exact production steps are slightly dissimilar between organizations.
[2]Cl: Chloride ions.
[3]$BaCO_3$: Barium carbonate.
[4]F: Fluorine.
[5]Bio-contaminant organisms, for example fungal (alternaria, fusarium and penicillium species) and bacterial (pseudomonas species).

Table 4. Anodizing processes for structural adhesion bonding of aluminum alloys (reproduced from Park et al. [46] with permission from Taylor & Francis).

Treatments	CAA [51]	PAA [55, 56]	BSAA [51]	PSA [53]
Oxide thickness (nm)	4000	200	3000	1500
Pore diameter (nm)	25	32	10	20–25
Cell wall thickness (nm)	10	18	10	–
Schematic view of oxide structure [nm] (non to scale)	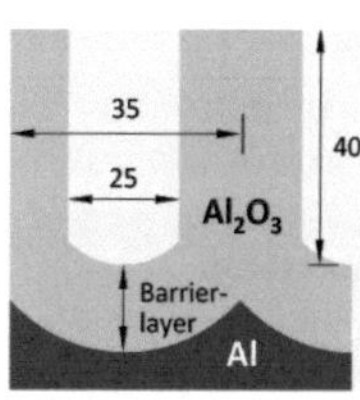	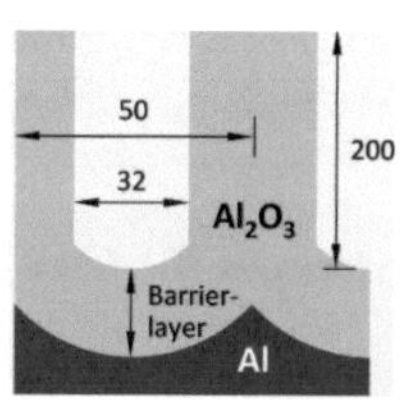	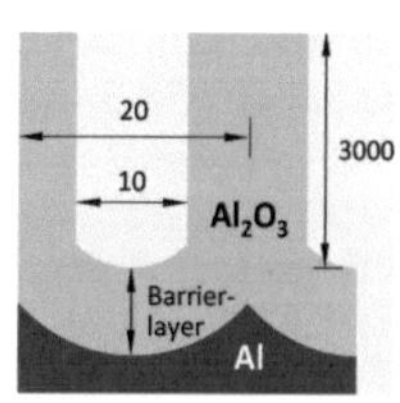	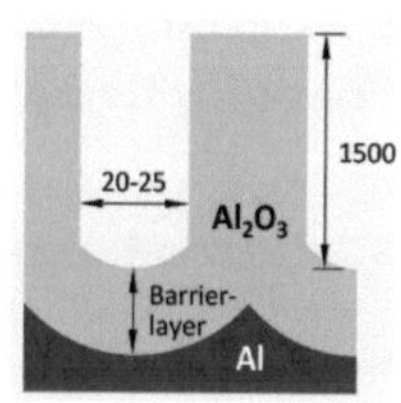

Table 5. Comparison of oxide morphology on 2024-T3 bare aluminum alloys (reproduced from Park et al. [46] with permission from Taylor & Francis).

thinner barrier layer [54, 55, 58]. The PAA oxide thickness is typically reported in the range from 200 to 400 nm with a much thinner barrier layer of about 10 nm [54, 55]. The physical comparisons between PAA and CAA oxide structures clearly represent the PAA oxide to have a much more open porous structure, which would be more easily penetrated by the subsequent organic bond primer, thereby drawing the organic polymer into the oxide structure to form a very strong interlocking interphase. The PAA oxide structure provides either equivalent or better durability results than the CAA oxide structure in the most experimental trails [59].

The BSAA process is usually carried out using a mixture of 5–10 wt.% boric acid (H_3BO_3) and 30–50 wt.% sulfuric acid (H_2SO_4) at 26.7 ± 2.2°C. This process was patented by Boeing as a direct replacement to the CAA process [51, 60]. It is well known that the CAA process produces a chromium mist that is hazardous to health if inhaled. The BSAA is an alternative that eliminates this concern and the need for mist control. The process standard, BAC 5632, involves deoxidizing with tri-acid solutions, consisting of sodium dichromate, sulfuric, and hydrofluoric acid (HF), followed by the application of boric and sulfuric acid anodizing. The parts are then dried in warm air at 75°C prior to bond primer application. The anodic film which is produced by the standard BSAA has relatively small pore diameter (10 nm) compared with the conventional CAA film (25 nm), as listed in **Table 5**. The anodic oxide structure from the BSAA has a paint adhesion that is equal, or superior, to the one formed on CAA [51, 60]. For this purpose, the BSAA process parameters have been modified by the research groups, for example [61]. As a result, the required surface topography and equivalent mechanical stability in strength and durability are only enhanced when the following process variations were instituted: electrolytic phosphoric acid deoxidizer (EPAD) [51]: anodizing bath temperature in the BSAA bath [51, 61] and additional post treatment using a PAD [51].

More recently, a variety of alternative chromate-free electrochemical treatments have been introduced in the context of corrosion protection and adhesive bonding of aluminum and its alloys. The new eco-efficient alternatives developed by Airbus include tartaric-sulfuric acid anodizing (TSA) and PSA. In particular, a significant step towards chromate-free has been achieved by PSA process for adhesively bonded joints. This process, which is utilized for adhesive bonded joints is usually carried out by using a mixture of 10 wt.% phosphoric acid (H_3PO_4) and 10 wt.% sulfuric acid (H_2SO_4) at temperatures ranging from 26 to 28°C [53]. This process is now ready for the qualification by Airbus. The standard process requires nitric acid deoxidizing prior to PSA treatment. The PSA-treated surface produces an oxide structure of about 1500 nm in thickness with somewhat narrow porous structures in the range from 20 to 25 nm in pore diameter [62]. The PSA process has a reduced process time (in typical, 23 min) and anodizing temperature (27°C), compared with the standard CAA [53]. This leads to an improvement in eco-efficiency by decreasing time and energy consumption and offers a capacity increase.

3.3. Lesson learned from serial production of GLARE structures

In-process QCs activities are essential if the fits, forms, functions and requirements designed into a part are to be consistently achieved. In general, the QC systems applied to the part production of GLARE structures should be established based on the company's own specifications, part requirements, engineering drawings. For this purpose, all available QC factors such as prescribed contractual requirements, available equipment, level of personnel training and documentation systems should be carefully considered. **Table 6** describes the specific lesson learned in part production and the corresponding practical solution.

Issues	Possible practical solutions
Fingerprints or fluid spots on aluminum sheets  [Sealed bags containing prepared rolled aluminum sheets prior to lay-up, source: CompositesWorld]	• The surface treated aluminum sheets should be stored in dust-free area, or be protected to prevent the further contamination. The sealed bags are typically utilized to protect the bond-primed surfaces of aluminum sheets prior to the lay-up process. • Only materials listed in the process specification shall be used in contact with the aluminum sheets inside the net trim line prior to cure. For this purpose, the consumable processing aid materials used in the parts production should be separated into two categories: (1) Contact-use materials: approved for use in direct contact with the anodized (or primed) surface of aluminum sheets. (2) Noncontact-use materials: approved for use as aids to processing but shall not contact the anodized (or primed) aluminum sheets inside the trim line prior to autoclave cure.
Autoclave pressure variations	• Pressure controller has to regulate the autoclave pressure to maintain a uniform pressure level throughout a cure cycle. In addition, the pressure reservoir shall be kept twice the autoclave pressure so as to operate the pneumatic valves and solenoid valves sufficiently.
Folds and kinks of aluminum sheet  [Roll out over prepreg layers, source: CompositesWorld]	• During the preparation of thin rolled aluminum sheets (0.2–0.5 mm), consisting of unrolling, cutting, surface treatments (i.e. anodizing and bond primering) and consequently rolling up, all aluminum sheets should be prepared without folding, or kinks. The damaged material shall not be used for part production. In the stage of lay-up process, the aluminum sheet should be rolled over the un-cured prepreg plies.
Folded prepreg during lay-up process	• As small regions of prepreg ply are sheared, the surrounding regions can begin to fold, or wrinkle because of the discontinuity in in-plane strain across the prepreg ply. If the deformation exceeds the limit angle of prepreg material, it is considered to be either fiber wrinkling, or bridging out of the material's tolerance. • Pre-cut materials shall be stored in flat, stress-free condition to prevent folding or damage. The damaged material shall not be used for part production.

Issues	Possible practical solutions
Porosities (or voids)  [NDT defect area with air enclosure, source: Fokker]	• The porosities (or voids) considerably degrade the static strength and fatigue life as a result of insufficient adhesion between aluminum–prepreg and prepreg–prepreg interfaces. Park et al. [6, 35] reported that the reduction in porosity content from 1.30% to 0.69% could account for 46.46% increase in the interlaminar shear strength. • High autoclave pressure (in typical. 11 bar) is generally applied to achieve the acceptable porosities (or voids) contents in GLARE parts.
Prepreg gap controls 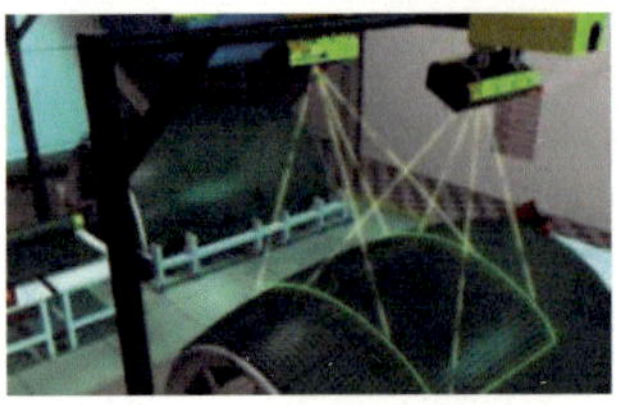 [LPS, source: Virtek]	• The intra-ply gap (i.e. butt-joint) and overlap splices is a common issue in the production of GLARE parts. It is apparent that the manufacturing–induced defects, such as delamination, or fiber missing result in the part thickness variation and the subsequent stress concentrations [63]. • The requirements of ply collation, such as fiber orientation, location and splice gap are normally specified on the engineering drawing, or corresponding process specification. A LPS with low-intensity laser beams may be utilized for precision controls of ply location, especially on tapered or contoured parts.
Spring-back after curing process	• One key challenge in GLARE part production is to fabricate the sound part within tight dimensional tolerances. This issue has been particularly found in the integrated GLARE fuselage panels, for example double curved panels manufactured by SFT process. Orthotropic thermal and chemical properties in combination with autoclave production parameters, such as cure temperature and compaction pressure, are detrimental to achieve this goal. • Being able to predict the changes in part configuration allows to design curing tool geometries that already compensate for the undesired change in curvature [64]. This approach can lead to a significant cost reduction for the curing tool-design.
Disturbed resin squeeze out after curing	• Insignificant resin bleeding, or squeeze-out in the transverse direction is occasionally observed between two aluminum sheets. This result is attributed to either the loss of vacuum pressure applied consistently, or the absence of curing reaction for epoxy-matrix. • At the same time, high compaction pressure is required to compress the materials and squeeze out excess resin.

Table 6. Lesson learned from production of GLARE structures.

4. Conclusions

The new hybrid material FML has been successfully applied to the commercial aircraft structures by offering weight savings of 10% compared with conventional aluminum and its alloys, together with benefits that include high tensile strength and better F&DT characteristics and high level of fiber safety. A large number of literatures on the practical applications demonstrates that the material properties of FMLs and their additional interlinked advantages make them the ideal option for thin-walled fuselage shells of next single aisle aircrafts. This chapter dealt with the details of technological developments with ongoing research efforts to understand the material property behaviors of FMLs, especially static strength, F&DT properties and long-term durability. In addition, two prediction methods of MVF and CLT have been introduced to predict the corresponding static properties of FMLs respect to the different lay-up patterns. However, to compete with the typical materials used in aerospace engineering, additional efforts should be directed towards producing consistently sound FML structures at affordable costs and ensuring the stringent quality controls for compliance with structural integrity. Recently, the FML manufacturers have continued to make a substantial progress in production technology, which allows for enabling FMLs in high-volume production rates and increasing affordability for aerospace industry. In addition to the consideration of each constituent material's properties, a strong interfacial bonding between metal sheets and composite layers is one of the key factors for the improvement in joint strength and long-term durability of FML structures. Therefore, a proper surface treatment on the metallic substrate is prerequisite for achieving long-term service capability through more efficient processing in production.

Author details

Sang Yoon Park[1]* and Won Jong Choi[2]

*Address all correspondence to: hanavia@empas.com

1 Hyundai Automotive Research and Development Division, Hwaseong-Si, Gyeonggi-Do, South Korea

2 Department of Materials Engineering, Korea Aerospace University, Koyang-city, Gyeonggi-Do, South Korea

References

[1] Park Y, O'Kelly ME. Examination of cost-efficient aircraft fleets using empirical operation data in US aviation markets. Journal of Air Transport Management. 2018;**69**:224-234. DOI: 10.1016/j.jairtraman.2017.02.002

[2] Alderliesten RC, Homan J. Fatigue and damage tolerance issues of GLARE in aircraft structures. International Journal of Fatigue. 2006;**28**:1116-1123. DOI: 10.1016/j.ijfatigue.2006.02.015

[3] Fontain L. NDT on A380. In: 48th Annual Non–Destructive Testing (NDT) Forum; 20–22 September 2005; Orlando, Florida

[4] Beumler T. Fiber metal laminate structures–From laboratory to application. In: Invitation to an RAeS Lecture in Cooperation with the DGLR and VDI (Presented to the Royal Aerospace Society Hamburg Branch at Hamburg University of Applied Science); 29th October 2009; Hamburg, Germany

[5] Roebroeks GHJJ. Glare features. In: Vlot A, Gunnink JW, editors. Fibre Metal Laminates: An Introduction. The Netherlands: Kluwer Academic Publishers; 2001. pp. 23-37. DOI: 10.1007/978-94-010-0995-9_2

[6] Park SY, Choi WJ, Choi HS, Kwon H. Effects of surface pre-treatment and void content on GLARE laminate process characteristics. Journal of Materials Processing Technology. 2010;**210**:1008-1016. DOI: 10.1016/j.jmatprotec.2010.01.017

[7] Subbaramaiah R, Prusty BG, Pearce GMK, Lim SH, Thomson RS. Crashworthy response of fibre metal laminate top hat structures. Composite Structures. 2017;**160**:773-781. DOI: 10.1016/j.compstruct.2016.10.112

[8] Botelho EC, Almeida RS, Pardini LC, Rezende MC. Elastic properties of hygrothermally conditioned GLARE laminate. International Journal of Engineering Science. 2007;**45**: 163-172. DOI: 10.1016/j.ijengsci.2006.08.017

[9] Shim DJ, Alderliesten RC, Spearing SM, Burianek DA. Fatigue crack growth prediction in GLARE hybrid laminates. Composites Science and Technology. 2003;**63**:1759-1767. DOI: 10.1016/S0266-3538(03)00082-4

[10] Müller B, Hagenbeek M, Sinke J. Thermal cycling of (heated) fibre metal laminates. Composite Structures. 2016;**152**:106-116. DOI: 10.1016/j.compstruct.2016.05.020

[11] Sinmazçelik T, Avcu E, Özgür Bora M, Çoban O. A review: Fibre metal laminates, background, bonding types and applied test methods. Materials & Design. 2011;**32**:3671-3685. DOI: 10.1016/j.matdes.2011.03.011

[12] Rans CD. Bolted joints in glass reinforced aluminium (GLARE) and other hybrid fibre metal laminates (FML). In: Camanho P, Tong L, editors. Composite Joints and Connections: Principles, Modelling and Testing. Oxford: Woodhead Publishing Limited; 2011. pp. 35-76. DOI: 10.1533/9780857094926.1.35

[13] Assler H, Telgkamp J. Design of aircraft structures under special consideration of NDT. In: 9th European Conference on NDT; 25–29 September 2006; Berlin, Germany

[14] Asundi A, Choi AYN. Fiber metal laminates: An advanced material for future aircraft. Journal of Materials Processing Technology. 1997;**63**:384-394. DOI: 10.1016/S0924-0136(96) 02652-0

[15] Vogelesang LB, Vlot A. Development of fibre metal laminates for advanced. Journal of Materials. Processing Technology. 2000;**103**:1-5. DOI: 10.1016/S0924-0136(00)00411-8

[16] Ye L, Lu Y, Su Z, Meng G. Functionalized composite structures for new generation airframes: A review. Composites Science and Technology. 2005;**65**:1436-1446. DOI: 10.1016/j.compscitech.2004.12.015

[17] Schut J, Alderliesten RC. Delamination growth rate at low and elevated temperatures in GLARE. In: 25th International Congress of the Aeronautical Sciences; 3-8 September 2006; Hamburg, Germany. pp. 1-7

[18] Botelho EC, Silva RA, Pardini LC, Rezende MC. A review on the development and properties of continuous fiber/epoxy/aluminum hybrid composites for aircraft structures. Materials Research. 2006;**9**:247-256. DOI: 10.1590/S1516-14392006000300002

[19] Hamill L, Hofmann DC, Nutt S. Galvanic corrosion and mechanical behavior of fiber metal laminates of metallic glass and carbon fiber composites. Advanced Engineering Materials. 2017;**20**:1700-1711. DOI: 10.1002/adem.201700711

[20] Wang WX, Takao Y, Matsubara T. Galvanic corrosion-resistant carbon fiber metal laminates. In: 16th International Conference on Composite Materials; 8–13 July 2007; Kyoto, Japan. pp. 1-10

[21] Pan Y, Wu X, Zh H, Wu G, Sun S, Ye H, Zhang Z. A new approach to enhancing interlaminar strength and galvanic corrosion resistance of CFRP/Mg laminates. Composites Part A: Applied Science and Manufacturing. 2018;**105**:78-86. DOI: 10.1016/j.compositesa.2017.11.009

[22] Vlot A, Vogelesang LB, de Vries TJ. Towards application of fibre metal laminates in large aircraft. Aircraft Engineering and Aerospace Technology. 1999;**71**:558-570. DOI: 10.1108/00022669910303711

[23] Wittenberg TC, van Baten TJ, de Boer A. Design of fiber metal laminate shear panels for ultra-high capacity aircraft. Aircraft Design. 2001;**4**:99-113. DOI: 10.1016/S1369-8869(01)00003-9

[24] Along the Bond Line–Groundbreaking Aircraft Structures [Internet]. Available from: (http://www.fokker.com/sites/default/files/media/Files/Brochures/Fokker_Glare.pdf) [Accessed: April 14, 2018]

[25] Campilho RDSG, da Silva LFM, Banea MD. Adhesive bonding of polymer composites to lightweight metals. In: Amancio-Filho ST, Blaga LA, editors. Joining of Polymer-Metal Hybrid Structures: Principles and Applications. New York: John Wiley & Sons, Inc; 2017. pp. 29-59. DOI: 10.1002/9781119429807.ch2

[26] Wu G, Yang JM. The mechanical behavior of GLARE laminates for aircraft structures. The Journal of the Minerals, Metals & Materials Society. 2005;**57**:72-79. DOI: 10.1007/s11837-005-0067-4

[27] Spronk SWF, Şen I, Alderliesten RC. Predicting fatigue crack initiation in fibre metal laminates based on metal fatigue test data. International Journal of Fatigue. 2015;**70**:428-439. DOI: 10.1016/j.ijfatigue.2014.07.004

[28] Vogelesang LB, Schijve J, Fredell R. Fibre-metal laminates: Damage tolerant aerospace materials. In: Demaid A, de Wit JHW, editors. Case Studies in Manufacturing with Advanced Materials, Vol. 2. Amsterdam: Elsevier; 1995. pp. 253-271. DOI: 10.1016/0925-8388(96)80048-X

[29] Hagenbeek M, van Hengel C, Bosker OJ, Vermeeren CAJR. Static properties of fibre metal laminates. Applied Composite Materials. 2003;**10**:207-222. DOI: 10.1023/A:1025569316827

[30] Vogelesang B, Gunnink JW, Roebroeks GHJJ, Muè Ller RPG. Toward the supportable and durable aircraft fuselage structure. In: Proceedings of the 18th Symposium of the International Committee on Aeronautical Fatigue; 3–5 May 1995; Melbourne, Australia. pp. 257-272

[31] Cortes P, Cantwell WJ. The prediction of tensile failure in titanium-based thermoplastic fibre–metal laminates. Composites Science and Technology. 2006;**66**:2306-2316. DOI: 10.1016/j.compscitech.2005.11.031

[32] Cortés P, Cantwell WJ. Fracture properties of a fiber-metal laminates based on magnesium alloy. Journal of Materials Science 2004;**39**:1081-1083. DOI: 10.1016/j.compositesb.2005.06.002

[33] Alderliesten RC, Hagenbeek M, Homan JJ, Hooijmeijer PA, de Vries TJ, Vermeeren CAJR. Fatigue and damage tolerance of GLARE. Applied Composite Materials. 2003;**10**:223-242. DOI: 10.1023/A:1025537818644

[34] Hooijmeijer PA. Burn-through and lightning strike. In: Vlot A, Gunnink JW, editors. Fibre Metal Laminates: An Introduction. The Netherlands: Kluwer Academic Publishers; 2001. pp. 399-408. DOI: 10.1007/978-94-010-0995-9_26

[35] Park SY, Choi WJ, Choi HS. The effects of void contents on the long-term hygrothermal behaviors of glass/epoxy and GLARE laminates. Composite Structures. 2010;**92**:18-24. DOI: 10.1016/j.compstruct.2009.06.006

[36] Remmers JJC, de Borst R. Delamination buckling of fibre–metal laminates. Composites Science and Technology. 2001;**61**:2207-2213. DOI: 10.1016/S0266-3538(01)00114-2

[37] Frizzell RM, McCarthy CT, McCarthy MA. A comparative study of the pin-bearing responses of two glass-based fibre metal laminates. Composites Science and Technology. 2008;**68**:3314-3321. DOI: 10.1016/j.compscitech.2008.08.021

[38] Şen I, Alderliesten RC, Benedictus R. Lay-up optimisation of fibre metal laminates based on fatigue crack propagation and residual strength. Composite Structures. 2015;**124**:77-87. DOI: 10.1016/j.compstruct.2014.12.060

[39] Sadighi M, Alderliesten RC, Benedictus R. Impact resistance of fiber-metal laminates: A review. International Journal of Impact Engineering. 2012;**49**:77-90. DOI: 10.1016/j.ijimpeng.2012.05.006

[40] Wischmann G. In: Wissenschaftliches Festkolloquium; 2006. Available online at www.iff.uni–stuttgart.de

[41] Airbus Customers Benefit from Fiber Metal Laminates [Internet]. Available from: (https://www.scribd.com/document/35637377/ILA06-Praesentat-AIRBUS-Juergen-Pleitner) [Accessed: April 14, 2018]

[42] Muchiri AK. Maintenance Planning Optimisation for the Boeing 737 Next Generation [Thesis]. Delft, The Netherlands: Delft University of Technology; 2002

[43] Rans C, Morinière F, Rodi R, Alderliesten R, Benedictus R. Fatigue behavior of fiber/metal laminate panels containing internal carbon tear straps. Journal of Aircraft. 2011;**48**:2122-2129. DOI: 10.2514/1.C031475

[44] Beumler T. Flying GLARE [Thesis]. Delft, The Netherlands: Delft University of Technology; 2004

[45] Alderliesten R, Rans C, Benedictus R. The applicability of magnesium based fibre metal laminates in aerospace structures. Composites Science and Technology 2008;**68**:2983-2993. DOI: 10.1016/j.compscitech.2008.06.017

[46] Park SY, Choi WJ, Choi HS, Kwon H, Kim SH. Recent trends in surface treatment technologies for airframe adhesive bonding processing: A review (1995–2008). The Journal of Adhesion. 2010;**86**:192-221. DOI: 10.1080/00218460903418345

[47] Zhang X, Ma Q, Dai Y, Hu F, Liu G, Xu Z, Wei G, Xu T, Zeng Q, Xie W. Effects of surface treatments and bonding types on the interfacial behavior of fiber metal laminate based on magnesium alloy. Applied Surface Science. 2018;**427**:897-906. DOI: 10.1016/j.apsusc.2017.09.024

[48] Li X, Zhang X, Zhang H, Yang JL, Nia AB, Chai GB. Mechanical behaviors of Ti/CFRP/Ti laminates with different surface treatments of titanium sheets. Composite Structures. 2017;**163**:21-31. DOI: 10.1016/j.compstruct.2016.12.033

[49] Pan YC, Wu GQ, Huang Z, Li MY, Ji SD, Zhang ZK. Effects of surface pre-treatments on mode I and mode II interlaminar strength of CFRP/Mg laminates. Surface and Coatings Technology. 2017;**319**:309-317. DOI: 10.1016/j.surfcoat.2017.04.010

[50] Aghamohammadi H, Abbandanak SNH, Eslami-Farsani R, Siadati SMH. Effects of various aluminum surface treatments on the basalt fiber metal laminates interlaminar adhesion. International Journal of Adhesion and Adhesives. 2018;**84**:184-193. DOI: 10.1016/j.ijadhadh.2018.03.005

[51] Critchlow GW, Yendall KA, Bahrani D, Quinn A, Andrews F. Strategies for the replacement of chromic acid anodising for the structural bonding of aluminium alloys. International Journal of Adhesion and Adhesives. 2006;**26**:419-453. DOI: 10.1016/j.ijadhadh.2005.07.001

[52] Bjørgum A, Lapique F, Walmsley J, Redford K. Anodising as pre-treatment for structural bonding. International Journal of Adhesion and Adhesives. 2003;**23**:401-412. DOI: 10.1016/S0143-7496(03)00071-X

[53] Matz CW, Hilling B, Kelm W, Kock E. Phosphoric Sulfuric Acid Anodizing (PSA) – A Heavy Metal Free Alternative for High Quality Surface Pretreatment of Aluminum. In: 83rd Meeting of the AGARD Structures and Materials Panel; 4–5 September 1996; Florence, Italy

[54] Oosting R. Toward a New Durable and Environmentally Compliant Adhesive Bonding Process for Aluminum Alloys [Thesis]. Delft, The Netherlands: Delft University of Technology; 1995

[55] Brockmann W, Hennemann OD, Kollek H. Surface properties and adhesion in bonding aluminium alloys by adhesives. International Journal of Adhesion and Adhesives. 1982;**2**: 33-40. DOI: 10.1016/0143-7496(82)90064-1

[56] Smith T, Lindberg G. Surface tools for automated non-destructive inspection of contamination. Surface Technology. 1979;**9**:1-29. DOI: 10.1016/0376-4583(79)90080-3

[57] Everett RA, Johnson WS. Repeatability of mixed-mode adhesive debonding. In: Johnson WS, editors. Delamination and Debonding of Materials. Philadelphia: ASTM International; 1883. pp. 267-281. DOI: 10.1520/STP36309S

[58] Venables JD. Adhesion and durability of metal-polymer bonds. Journal of Materials Science. 1984;**19**:2431-2453. DOI: 10.1007/BF00550796

[59] Bishopp J. Surface pretreatment for structural bonding. In: Cognard P, editor. Handbook of Adhesives and Sealants. Vol. 1. Amsterdam: Elsevier; 2005. pp. 163-214. DOI: 10.1016/S1874-5695(02)80005-7

[60] Zhang JS, Zhao XH, Zuo Y, Xiong JP. The bonding strength and corrosion resistance of aluminum alloy by anodizing treatment in a phosphoric acid modified boric acid/sulfuric acid bath. Surface and Coatings Technology. 2008;**202**:3149-3156. DOI: 10.1016/j.surfcoat.2007.10.041

[61] Domingues L, Fernandes JCS, Cunha Belo MD, Ferreira MGS, Guerra-Rosa L. Anodising of Al 2024-T3 in a modified sulphuric acid/boric acid bath for aeronautical applications. Corrosion Science. 2003;**45**:149-160. DOI: 10.1016/S0010-938X(02)00082-3

[62] Velterop L. Phosphoric Sulphuric Acid Anodising: An Alternative for Chromic Acid Anodising in Aerospace Applications. Amsterdam, The Netherlands: National Aerospace Laboratory (NLR). Report number: NLR-TP-2003–210; 2003

[63] Nardi D, Abouhamzeh M, Leonard R, Sinke J. Detection and evaluation of prepreg gaps and overlaps in GLARE laminates. Applied Composite Materials. DOI: 10.1007/s10443-018-9679-z

[64] Abouhamzeh M, Sinke J, Benedictus R. Investigation of curing effects on distortion of fibre metal laminates. Composite Structures. 2015;**122**:546-552. DOI: 10.1016/j.compstruct.2014.12.019

Chapter 6

Fiber-Metal Laminate Panels Reinforced with Metal Pins

Ruham Pablo Reis, Iaroslav Skhabovskyi,
Alberto Lima Santos, Leonardo Sanches,
Edson Cocchieri Botelho and Américo Scotti

Additional information is available at the end of the chapter

http://dx.doi.org/10.5772/intechopen.78405

Abstract

Fiber-metal laminates (FMLs) are key to modern composite structures and metal-composite coupling is crucial to improve their effectiveness. Cold-metal transfer (CMT) PIN welding, in correlated efforts, has been successfully explored as a metal-composite hybrid joining approach. This work proposes a novel development on FMLs, which consists of introducing metal pins welded by CMT PIN for anchoring their metal and composite layers together. Thus, miniaturized FML panels with different pin deposition spacing and patterns are evaluated with emphasis in drop-weight testing followed by buckling and by means of Iosipescu shear test as complement. They are also subjected to cosmetic and preliminary modal analyses. Besides not adding significant weight, the pins does not make the panels more brittle and their distribution does not imply significant effect in the capacity that the panels have to dissipate impact. The panels with pins exhibit a less catastrophic trend, indicating damage tolerance improvement as significantly higher loads at longer axial displacements in buckling test after impact are achieved. The anchoring effect of the pins is confirmed throughout the shear test results. The pins also significantly increase the damping factor of the panels and the changes in their metal surfaces by the CMT PIN process are considered as irrelevant.

Keywords: fiber-metal laminates, composite structures, adhesion, anchoring, CMT PIN

1. Introduction

The demand for high-performance-lightweight structures continues to stimulated advances in fiber- metal laminates (FMLs). Cortes and Cantwell [1] describe FMLs as hybrid composite structures based on thin sheets of metal alloys and plies of fiber-reinforced polymeric materials. Sinmazçelik et al. [2] justify this combination by considering that metals are isotropic materials, have high strength and impact resistance and are easy to repair, while full composites generally present high strength and stiffness and excellent fatigue characteristics. According to Salve et al. [3], FMLs take positive characteristics from both metals and fiber-reinforced composites, resulting in superior mechanical properties compared to conventional lamina processed from fiber-reinforced composites or monolithic metals. Despite exhibiting low weights, according to Seydel and Chang [4], composite materials are very susceptive to impact damage. On the other hand, metals are in general heavier, but might have excellent impact resistance. Thus, very light and impact resistant panels might be produced by combining the low volumetric mass density of a fiber-reinforced polymer with the high impact resistance of ductile metals. As stated by Vlot [5], impact damage zones of FMLs are smaller than those found in fiber-reinforced composites alone. In summary, FMLs are characterized by a balance of low structural weight and high strength and stiffness, when compared with metals.

The adhesion between metal and composite is clearly a crucial factor to improve the effectiveness of FMLs and several techniques accomplish it by changing the surface characteristics of the metal part [2]. In this line, hybrid joining approaches have been developed, but specifically aiming at improving the performance of composite-metal joints [6]. These approaches essentially rely on macro-scale metal anchorages on the metal surface to interlock the layers of fiber-reinforced polymer and add strength to the adhesive bond. One of the most recent techniques applies an array of pins deposited on the metal part, before joining, as anchorages (mechanical interlocking) for the composite part. Cold-metal transfer (CMT) PIN, an arc-welding-based process developed and commercialized by Fronius (Austria), would be a relatively low-cost option to be used. The CMT PIN was created from its parent process called CMT, in which metal wire is continuously fed and an electric arc (protected by gas) is open between the wire tip and the material to be welded, as in all gas metal arc welding (GMAW) techniques. But in difference to common GMAW, CMT uses a proprietary torch, which periodically reverses the advancing electrode wire when it touches the weld pool, allowing a smooth transfer of molten material into it (in this instant, current is also considerably reduced, with no short-circuit current peaks) without spattering (loss of material). The heat transferred to the work piece is very low, causing just minor metallurgical changes as well as low levels of distortion and discoloration/oxidation (including in the reverse side) in the base metal. The CMT PIN embarks a program for pin depositions. Following short-circuiting, it causes the wire to stay resting against the weld pool (with no arc) allowing time for cooling and consequent welding of the wire to the base metal. Shortly afterwards, a low current is forced to heat and soften the welded wire near its mid length (between the torch and the work piece), but not enough to break it apart and open the arc. With the reversing motion started, the soft wire undergoes a tensile force and is broken, leaving a pin formed on the base metal. By changing deposition parameters, pins of different sizes and geometries can be formed [7]. Ucsnik et al. [8] have shown that pins deposited by CMT PIN on the metal side improved the strength of metal-composite joints. According to Graham et al. [9], adhesive bonding with CMT PIN anchorage was able to consistently outperform adhesive bonded only specimens in terms of strength and energy absorption at quasi-static and high loading rates, as well as in terms of damage tolerance after impact, environmental durability and mechanical fatigue performance.

The capacity that the CMT PIN process seems to have to enhance the performance of composite-metal hybrid joints might be extended to FMLs. Arrays of pins on the metal sides could perform as anchorages for the composite sides throughout the panels, contributing to the performance along with the adhesive bonding between the composite and metal sheets. This approach has been recently investigated for the first time and the initial results are promising [10], despite the first pinned panels, due to manufacturing issues related to the number of prepreg layers, have exhibited significant variation in thickness. The first results indicate that the pins have potential to improve damage tolerance of fiber-metal laminate panels (FMLPs). Thus, the objective of this present work was to assess the application of metal pins deposited by CMT PIN on metal surfaces used as layers of FMLPs, yet controlling the thickness and the number of prepreg layers. The methodological approach includes comparing small-sized FMLPs with different pin deposition patterns and spans to a reference (without pins) FMLP, in terms of energy dissipation during drop-weight testing, impact damage characterization and buckling test after impact. Iosipescu shear test, modal analysis and cosmetic characterization were also carried out.

2. Methodological approach

2.1. Manufacturing of the panels

The small-sized FMLPs reinforced with metal pins were produced as illustrated in **Figure 1**. The reference FMLPs (without pins) were produced in the same way, with exception of the pin deposition step. The metals parts were composed of AISI 430 stainless steel sheets (350 × 80 × 0.4 mm). For pin deposition, a Fronius TransPuls Synergic 5000 power source was used, connected to a VR7000-CMT wire feeder, a PullMig CMT torch and a RCU 5000i remote control unit, with DB0875 data base selected (synergic line CrNi 19 9 PIN). A welding robot was used to move the torch according to the deposition pattern. An AWS ER309L filler wire with nominal diameter of 1 mm (verified value of 0.98 ± 0.00 mm) was used as pin material and Ar with 4% (verified as 3.7%) of CO_2 gas at a flow rate of 8 L/min was employed to protect the pin deposition. Fronius Contec MD® contact tips were employed, as recommended for CMT PIN. The torch was always kept perpendicular to the metal parts. The metal surfaces were all homogeneously wiped with acetone wetted cloth. Other input parameters, such as contact-tip to work-piece distance (5 mm), metal sheet temperature (room temperature ≈ 27°C) and pin deposition sequence were remained unchanged. The CMT PIN process was then parameterized to get minimum-height (considering the thickness of the FMLs) and small ball-head pins. Ucsnik et al. [8] had shown that the ball-head shaped pins present potential in metal-composite hybrid joining. Pins of 2.50 ± 0.06 mm in height and 1.40 ± 0.03 mm in head diameter were produced, resulting in an average pin weight of 0.023 ± 0.002 g.

The pin deposition process was monitored by sampling the electrical current and voltage data, as exemplified in **Figure 2**. The pin deposition cycle (T) was 5.73 s, consequence of the robot motion between two consecutive deposition points, which was not optimized in this work. As seen, the CMT PIN process works basically by controlling the levels of current applied to the electrode-wire and the dwell time in each level. First, the voltage is at its open-circuit level ready for arc starting. Then, the electrical arc is struck (by short-circuiting the electrode-wire to the base metal) and its tip and base metal below are melted (t_{arc} = 0.04 s). Subsequently, after the electrode-wire forward movement and contact with the base metal, the arc is extinguished and the welding between the electrode-wire and the base metal takes place (t_{weld} = 0.02 s).

Finally, the welded electrode-wire undergoes heating, softening and breaking apart at its mid length range by a combination of current flow and electrode-wire backward movement (t_{rup} = 0.36 s).

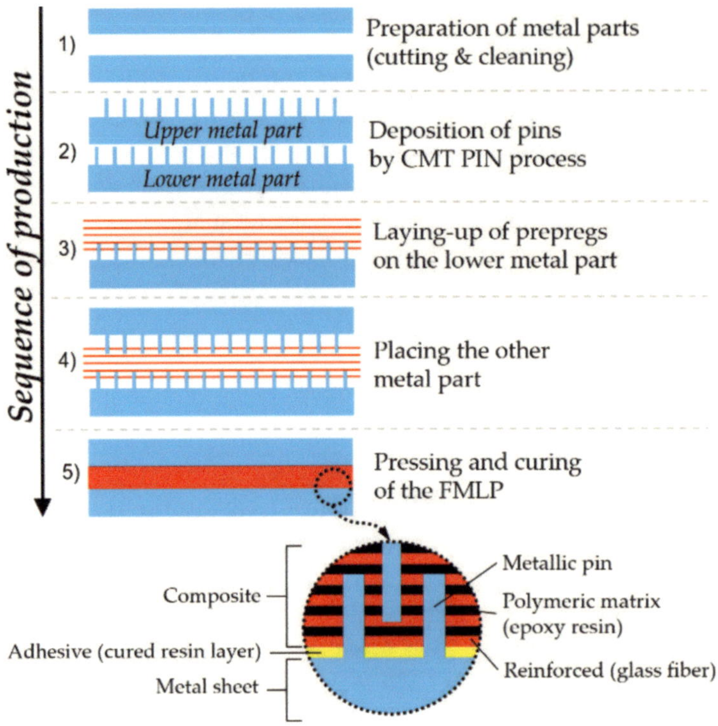

Figure 1. Sequence of production of small-sized FMLPs reinforced with metal pins welded by cold-metal transfer (CMT) PIN process.

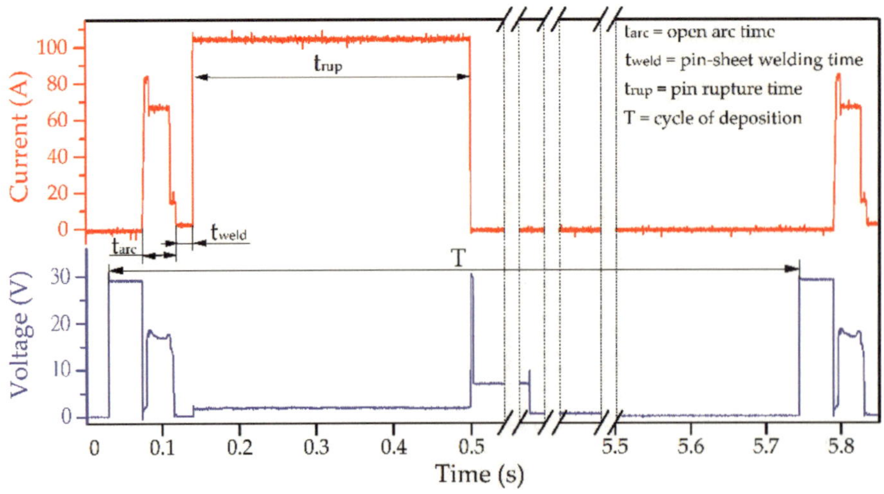

Figure 2. Cycle of pin deposition based on the electrical current and voltage data.

After CMT PIN processing, all metal parts (with and without pins) went through ultrasonic cleaning by immersion in acetone for 8 minutes to remove any process-related or not elements that could impair the adherence of the composite parts (**Figure 3**). For composite lamination, prepregs made of glass fiber (8-Harness Satin Weave) and epoxy resin from Hexcel Corporation (product data 7781-38"-F155) were employed. Each layer of prepreg was cut with dimensions of 360 × 90 mm (in excess of 5 mm measured from each of the metal sheet edges). The prepreg layers were stacked on top of the lower metal parts (with or without pins), always aligning the warp yarns with their length and, consequently, the filling yarns with their width. Next, the upper metal parts were placed aligned to the prepreg layers and lower metal parts. All prepreg handling was executed in a white room, avoiding to the maximum any contamination. All small-sized panels were wrapped with a thin film of release agent (polyamide) and then processed in groups of three in a CARVER® CMG100H-15-C hot-curing press according to the curing cycle displayed in **Figure 4**. After curing, the composite material exciding the edges of each panel was removed by means of band sawing followed by belt sanding, ending up in panels of 350 × 80 × 4 mm.

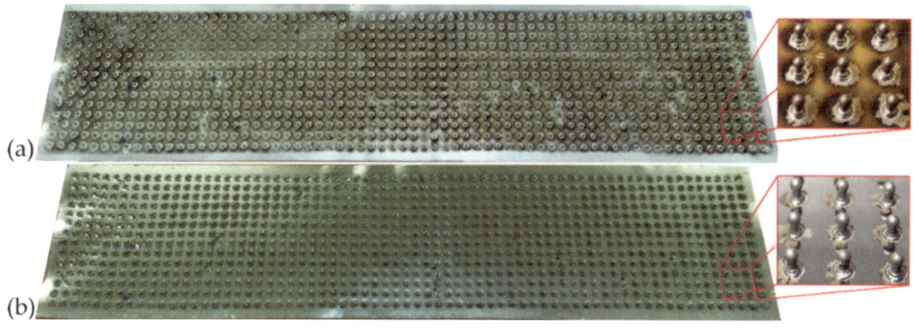

Figure 3. Example of metal sheet after pin deposition, where: (a) before; and (b) after cleaning.

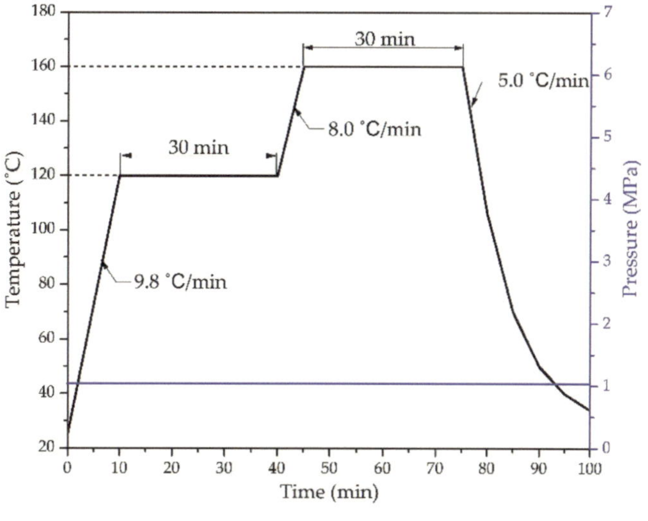

Figure 4. Curing cycle used for the fabrication of the small-sized FMLPs.

Two basic types of small-sized FMLPs were made (**Figure 5**): conventional FMLPs (without pins), i.e., metal-composite-metal (M-C-M); and FMLPs with pins, i.e., pined metal-composite-pined metal (MPin-C-MPin). The MPin-C-MPin type was produced using a combination of two levels of pin separation (5 and 10 mm) and two deposition patterns (hexagonal and squared). The pins of the upper metal parts were deposited with an adequate displacement in relation to the pins of the lower metal parts, avoiding contact between the upper and lower pins and providing homogenous distribution. Each one of the five small-sized FMLPs types was produced twice for replication in the tests.

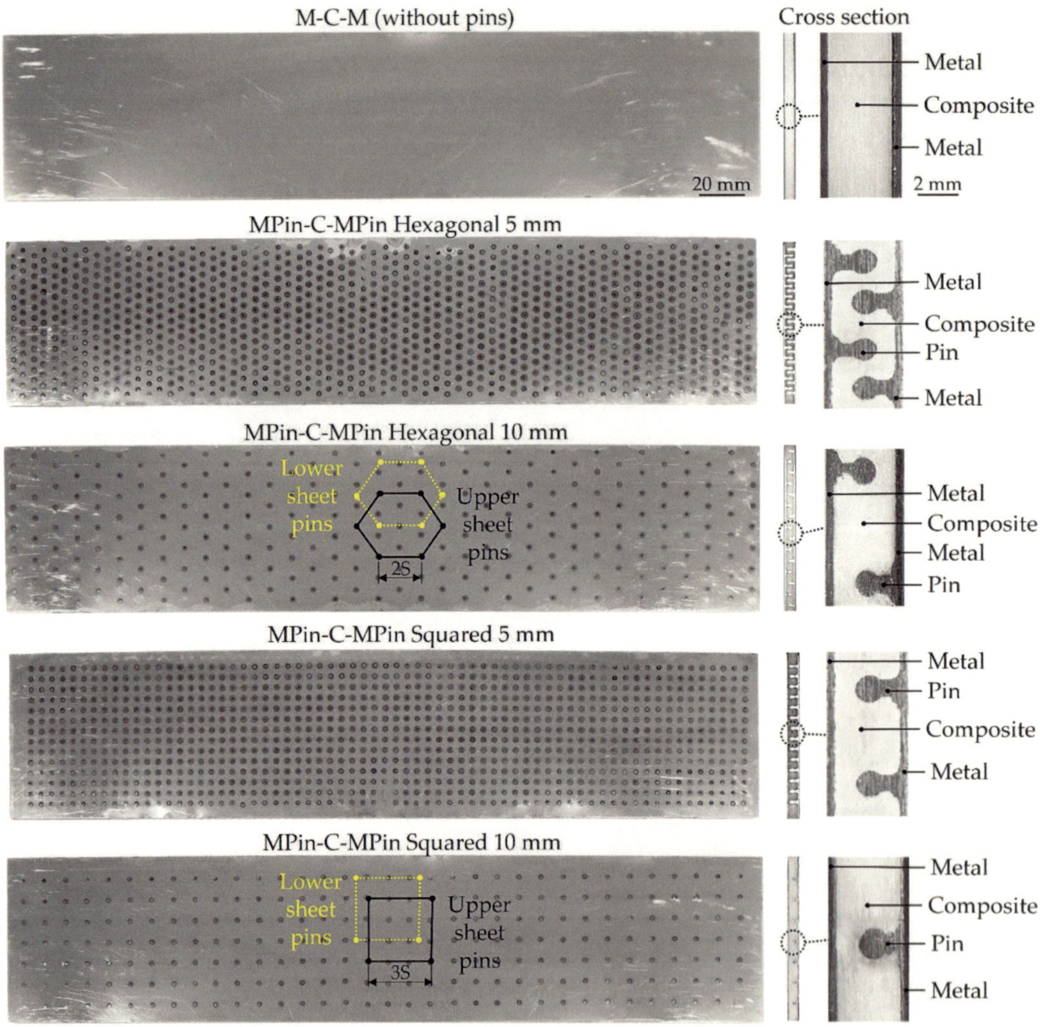

Figure 5. Types of small-sized panels fabricated with respective cross sections and schematic overlapping of pins on the upper and lower metal sheets, where M-C-M stands for metal-composite-metal, MPin-C-MPin for pined metal-composite-pined metal for hexagonal and squared pattern and S for distance separating the pins (panel width ≈ 80 mm; panel length ≈ 350 mm; the dot-like marks on the panels with pins are due to stainless steel heat-induced oxidation right under where the pins were deposited—This esthetic effect could be avoided with inert gas back purging).

Panel type	Pins per panel	Prepreg layers per panel	Actual thickness (mm)	Pin density (pin/cm²)	Mass density (g/cm³)
M-C-M (without pins)	0	13	3.90 ± 0.03	0	3.0
MPin-C-MPin hexagonal 5 mm	2001	13	4.04 ± 0.03	3.61	3.3
MPin-C-MPin hexagonal 10 mm	525	13	4.00 ± 0.07	0.94	3.0
MPin-C-MPin squared 5 mm	1976	13	4.08 ± 0.01	3.57	3.3
MPin-C-MPin squared 10 mm	518	13	3.96 ± 0.02	0.93	3.1

Table 1. Characteristics of the small-sized FMLPs.

As shown in **Table 1**, an attempt was made to keep the same thickness (approximately 4 mm) and prepreg layers per panel (13) for all FMLPs types. Pin density was estimated for each pertinent case, by considering the respective number of pins divided by the panel surface area (560 cm^2). However, concerning mass density, usually crucial for FMLPs, the presence of pins has only marginal effect, regardless of pin density. The overall average mass density of the FMLPs with pins was 3.2 g/cm^3, close to the case of M-C-M (without pins).

3. Non-destructive evaluation

3.1. Modal analysis

Modal analysis is a tool largely used to determine dynamic properties (natural frequencies, damping factors and vibration modes) of mechanical structures by imposing vibrations [11]. According to Rao [12], any movement of a flexible structure that repeats itself after a time interval is called vibration. This movement can be inferred instantaneously or continuously. The use of modal analysis is often applied due to the ease of implementation, relatively low cost, as well as being a nondestructive analysis.

According to Bolina et al. [13], the natural frequencies (fn) indicate the rate of free oscillation of the structure, after ceasing the force that caused its movement. In other words, it represents how much the structure vibrates when there is no longer a force applied to it. It is worth recalling that the value of the natural frequency of a structure depends on its stiffness and mass. In a structure several natural frequencies can be observed because it can vibrate freely (after being excited by a force) in several directions and modes. In practice, higher values of natural frequency indicate that elastic stresses are preponderant to inertial forces. Moreover, whenever a structure oscillates with a frequency equal to one its natural frequencies, a phenomenon called resonance occurs. The resonance implies high amplitudes of vibration, and can cause structural failures as, for example, in the breaking of a crystal glass of wine due to sound energy. In this case, when the frequency of vibration caused by the source (sound wave) coincides with the natural frequency of the crystal glass, the amplitude of oscillation of this body reaches high values, because the source progressively gives energy to the body, and the crystal glass might break if the strain extensions exceed the levels supported by the material. However, vibration levels during a resonance can be attenuated if there are dissipation mechanisms that have high damping factors present in the structure, for example, shock absorbers (holding tightly the crystal glass in one's hand avoid the resonance-induced breakage). A more damped structure at a certain natural frequency can attenuate vibration levels more quickly. The damping factors (ξn) represent the damping levels of a structure, which in turn are characterized for each of its natural frequencies.

By analyzing, for a given natural frequency (periodic movement) the relationship between the points that discretize the structure, one obtains the natural modes of vibration. That is, the modes of vibration determine the way the structure vibrates at a certain natural frequency. In practice, by knowing the dynamic properties of a structure, it is possible to determine how the vibration oscillations will be at different measurement positions. Therefore, the objective of applying experimental modal analysis in the FMLPs was to evaluate, through the respective natural frequencies (fn) and damping factors (ξn), whether the pins inside the FMLPs would be able to modify the dynamic properties of the panels. The modal forms of the FMLPs were not evaluated in this work.

According to Lundkvist [14], structural dynamic properties or modal parameters in practice are identified from the Frequency Response Function (FRF). To obtain the FRF, the data that are in time domain are transformed to frequency domain, using the Fourier transformation, as exemplified in **Figure 6**. At the same time, from the response of the system vibrating under a condition, the modal properties are determined. It can be said that from the identified modal properties, the dynamic temporal behavior of the structure is predicted under any excitation conditions. As shown in **Figure 6(d)**, a temporal response of the structure contains the participation of its modes of vibration, each mode of vibration contemplating a natural frequency, damping factor and modal form. In order to apply the modal analysis in the small-sized panels, the experimental bench shown in **Figure 7** was used.

Some details of the experimental bench assembly are given below:

- The small-sized FMLP is hung by means of a nylon line glued to the side edges of it, forming a 60 mm arrow, as shown in **Figure 8**. It is called a "free-free" boundary condition when there is no rigid support or fixation of the structure. In this way, FMLP would be, by this experimental setup, in a "free-free" boundary condition, which was chosen because it is relatively simple to perform and quite informative.
- The positions of analysis of the dynamic response of each panel were based on a mesh of 51 points, as also indicated in **Figure 8**.
- The piezoelectric accelerometer is a model 352C22 (S/N LW181487) from the manufacturer PCB Piezotronics®, fixed in a position of the small-sized FMLP by means of a specific wax, that allows the propagation of the vibrations between the panel and the acceleration.

- The shaker is a model K2007E01 from manufacturer Modal Shop Inc. with the function of programming time and frequency of vibration to induce in the object of analysis (panel) through a connection tube. In this work, the shaker was programmed to produce a white noise type signal (random signal with equal intensity at different frequencies) comprising the frequency range 0–800 Hz in order to find the dynamic properties of the structure. The choice of this frequency range was made from a preliminary study, in which the first four modes of vibration were found. This frequency band determines the composition of the white noise that will be exerted by the electromagnetic exciter.
- The load cell is a model 208C03 (S/N LW34448), glued to the panel at a given position, as shown in **Figure 8**.
- The data acquisition system is capable of obtaining and analyzing the accelerometer and load cell signals. This system was configured to obtain a frequency resolution (Δf) of 0.25 Hz and an analysis band between 0 (fmin) and 800 (fmax) Hz. The acquisition rate (temporal resolution) of the data is then defined automatically by the acquisition system.
- The software is from Brüel and Kjær, from the manufacturer Vibrant Technology®, for acquisition and post-processing to identify the dynamic properties parameters of the panels.

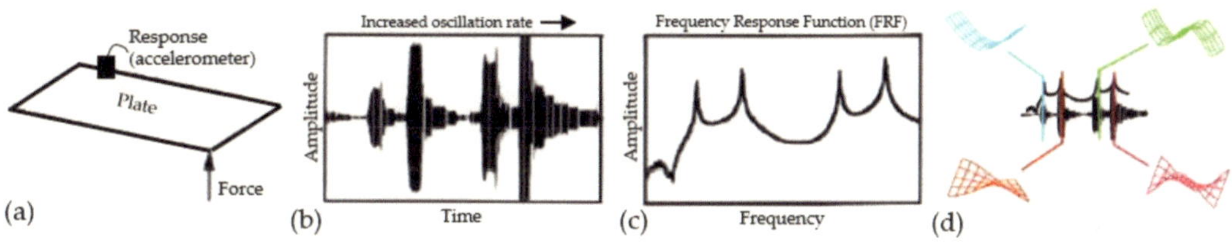

Figure 6. Illustration of a modal analysis, where: (a) experimental assembly; (b) response of the plate in the time domain; (c) response of the plate in the frequency domain (FRF); (d) overlap of the responses and example of the modal plate forms corresponding to each natural frequency (resonance) (after [15]).

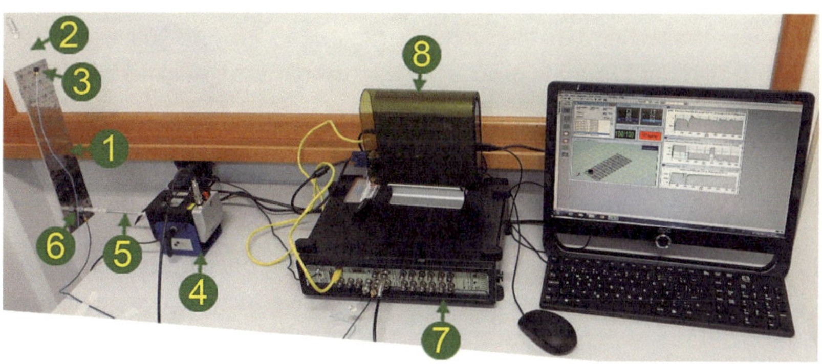

Figure 7. Experimental bench with the equipment used for perform modal analysis of the small-sized FMLPs, where: 1—panel (object of analysis); 2—nylon line glued to the edges of the FMLP; 3—accelerometer; 4—shaker; 5—stinger (connection tube) between the load cell and shaker; 6—load cell (bonded to the panel); 7—acquisition data system; 8—computer and software for data processing.

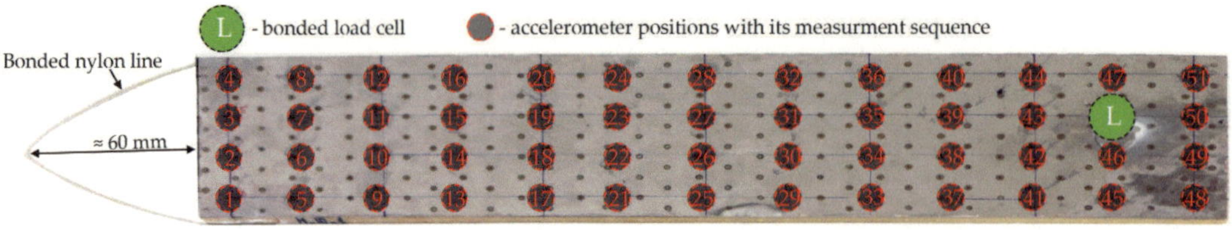

Figure 8. 51 mesh sequential point positioning of the accelerometer in a small-sized FMLP MPin-C-MPin hexagonal 10 mm type (≈350 × 80 × 4 mm).

Figure 9 shows the FRF curve typical of the amplitude and phase relationship (angular lag) along the frequency domain (0–800 Hz) between the output signals (measured acceleration) and input (applied force), from which the dynamic properties of the panels were determined. In this case, the phase between the responses of the accelerometer (acceleration) and the load cell (force) has a role to confirm if a resonant peak actually occurred along the frequency domain. Rao [12] mentions that resonance occurs when simultaneously a peak in the positive signal of FRF amplitude is observed and when the phase is near or passes through 90°. As shown in **Figure 9**, in the four resonance peaks (modes) the phase signals clearly changed their values (downwards), thus, by confirming that the resonance peaks actually occurred. The shape of the peaks (resonances) also indicates the level of damping, since the more "oval" peaks are associated with high damping effect. The Rational Fraction Polynomial (RFP) method [11, 16] was used to accurately determine the natural frequencies (fn) and damping factors (ξn) of each mode. The RFP method consists in adjusting a ratio of polynomials to FRF curves (amplitude and phase) obtained experimentally. For this purpose, a routine was devised in MATLAB® environment.

The dashed lines of **Figure 9** point to the appearance (typical for all panels analyzed) of the adjustment made by the RFP method in FRF (amplitude and phase) experimental curves using a 7th degree polynomial. With the adjustment performed in all positions, 51 modal parameters of each panel, namely the natural frequencies (fn) and damping factors (ξn), were determined. The solid lines of **Figure 9** show the FRF curves of the amplitude and phase evaluated in a given position of the panel (similar behavior for the other points of the mesh). It is possible to note the presence of four peaks (frequencies around 200, 480, 560 and 650 Hz) from the amplitude plot of **Figure 9**, indicating the resonant frequencies of the panel.

Parts (a) and (b) of the **Figure 10** present average results of four modes of the natural frequency (fn) and damping factor (ξn), respectively, measured at the 51 points of the measurement mesh. To facilitate the analysis, **Table 2** was organized to display the results of the four modes of vibration in terms of natural frequency and damping factor for all types of pinned small-sized panels in relation to the conventional one (without pins). Regarding the natural frequency, which represents the relationship between the stiffness and mass of a structure, it is possible to notice that the pinned panels did not have a significant variation of the natural frequencies of the vibration modes compared to the panels without pins. This implies that the pins have little effect on this dynamic property. Possibly, in spite of the mass increase provided by the deposition of the pins, the increase in stiffness takes place in a proportional way so that the natural frequency remains unchanged (in relation to the panel without pins). In contrast, the presence of the pins in the FMLPs significantly increased the damping factor of the FMLPs between two and five times (pinned panels with 10 mm spacing and hexagonal deposition pattern) depending on the vibration mode, compared to conventional panel (without pins). This was probably due to the fact

http://dx.doi.org/10.5772/intechopen.78405

that the pins, individually or group, acts as a power dissipation mechanism. Thus, the oscillation amplitudes are attenuated more quickly with the presence of the pins. This quality could be exploited in practice, for example, aiming at structures less susceptible to acoustic problems and to high amplitudes of vibration at low frequencies. Other patterns of deposition and/or pin density could be explored in the sense of improving acoustic insulation and damping characteristics of vibrations in structures based on metal-composite laminates.

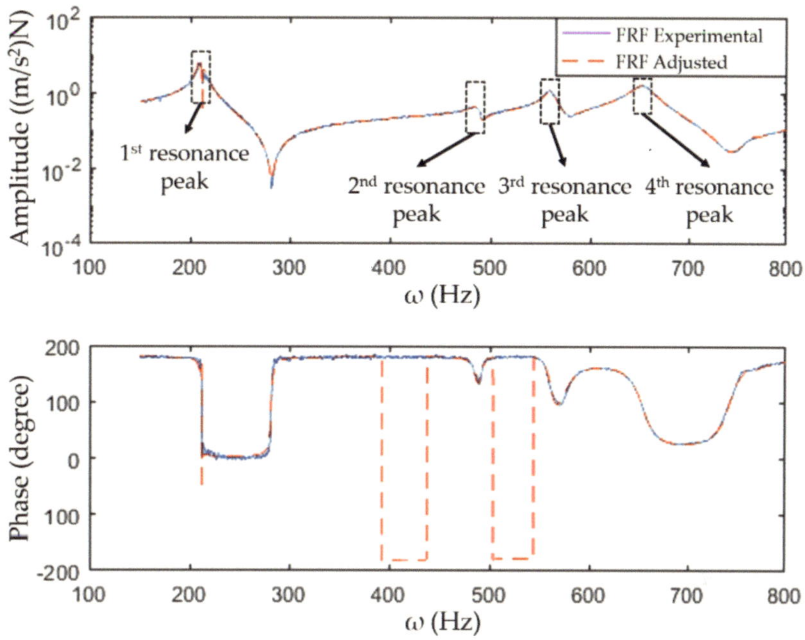

Figure 9. Amplitude (above) and phase (below) curves evaluated in a point of the MPin-C-MPin hexagonal 10 mm panel type (similar behavior for the other points), where solid lines are FRF and dashed lines are adjustment performed by the RFP method in the analysis band (ω) between 0 and 800 Hz (note that the adjustment overlapped the experimental curve for the amplitude case).

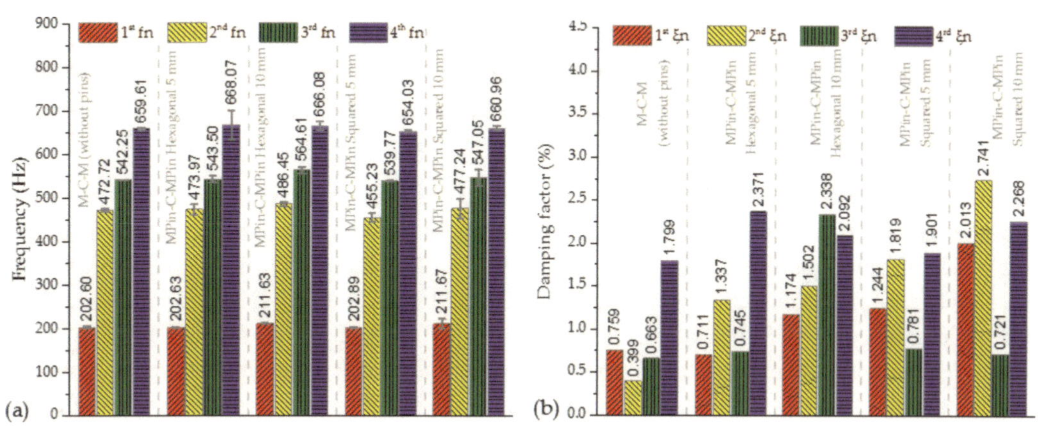

Figure 10. (a) Average results of the four natural frequency modes (fn) and (b) average results of the four modes of damping factors (ξn) obtained through modal analysis of the small-sized FMLPs.

Panel type	1st mode (%)		2nd mode (%)		3rd mode (%)		4th mode (%)	
	Δfn	Δξn	Δfn	Δξn	Δfn	Δξn	Δfn	Δξn
MPin-C-MPin hexagonal 5 mm	0.02	−6.32	4.46	54.68	0.14	63.90	4.48	165.22
MPin-C-MPin hexagonal 10 mm	0.26	235.09	2.90	276.44	−3.70	355.89	0.96	586.97
MPin-C-MPin squared 5 mm	0.23	12.37	4.12	252.64	−0.46	17.80	0.89	8.75
MPin-C-MPin squared 10 mm	1.28	31.80	0.98	16.29	−0.85	5.67	0.21	26.07

Table 2. Percentage change of vibration modes of the natural frequency (Δfn) and damping factor (Δξn) of the panels with pins in relation to the conventional panel (without pins).

3.2. Cosmetic characterization

In literature [17–19], the major part of Fiber-Metal Laminates (FMLs), due to their advantages, is used in modern aircraft parts, such as fuselage, wings, etc. FMLPs are used as manufactured or after paint application. In this context, the quality of the surface becomes a relevant factor. Thus, the objective of this section was to evaluate the influence of the deposition of the pins on the metal surface of the FMLP. A general evaluation of the cosmetic characteristics of the different FMLPs was made by measuring the waviness and roughness of the metal surfaces opposite to the pinned regions.

Figure 11 shows a small-sized FMLP reinforced with pins, where it is possible to perceive the dark marks on the metal surface, which represent the pins deposited on the opposite side. These marks are characterized by a localized deformation (peak and valley) and a heat-induced oxidation due to the welding of the pins (the cosmetic effect of the oxidation could be avoided with the application of purging with inert gas or reducing agent), that is, it is taken as a thermomechanical effect. The localized deformation is due to the operational mode of the CMT PIN, which deposits the pins by arc process (thermal effect), creating a small weld pool, but not reaching the opposite side. The tip of the electrode wire then penetrates into the weld pool, possibly reaching its bottom and pushing the surface of the metal sheet, deforming it (mechanical effect), characterizing the peak in the center of the marks (out of the panel). Then the current is turned off and the wire solidifies on the surface of the metal sheet. Next, current flows through the welded wire and metal sheet, promoting heating by Joule effect and wire softening, and a retraction movement of the wire breaks it apart, leaving behind a pin. It is believed that this retraction of the wire deforms the metal sheet, creating valleys concentric to the peak in the future panel. Concurrently, some expansion and contraction effects may also occur, but in a secondary way.

The measurements of surface waviness and roughness was performed with a Form Talysurf Intra profilometer from Taylor Hobson® with a resolution of 16 nm and range measurement of 1 mm. The measurement parameters were set as measuring speed equal to 0.25 mm/s, measuring length equal to 10 mm and ambient temperature of 20 ± 2°C (ABNT NBR ISO 1 [20]) and value of cut-off sample length equal to 0.8 mm (according to ISO 4288 [21]). Recalling that the cut-off value represents a segment of measurement length used in some representations of roughness measurements. It is important to note that, for the panels with pins, the tip of the profilometer was positioned to measure only some of the regions of marks left by the deposition of the pins crossing them radially. In the panels without pins, areas of the same size were chosen. For statistical purposes, three samples of surfaces for each type of panel, chosen randomly, were analyzed. **Figure 12(a)** and **(b)** show the typical waviness profiles of the surfaces of panels with and without pins, respectively, and illustrates the associated measured region. The quantification of the waviness was made from the maximum distances between peak and valley [ΔZ of **Figure 12(a)**] of each marking. It has been observed that the peak of the surface corrugation profile is located in the center of the pin deposited on the opposite surface of the metal sheet and the two minimum peaks appear around the pin, as shown in **Figure 12(a)**. Since the pins are made with the same parameters, the values of ΔZ are close, resulting in an average value of 26.92 ± 2.35 μm. For comparison purposes, the same criterion was applied in FMLPs without pins, resulting in average ΔZ of only 1.40 ± 0.40 μm.

The column graph of **Figure 13** shows the average (Ra), maximum (Rz) and total (Rt) roughness of the sampled regions of the panels. Recalling that Ra is the arithmetic mean

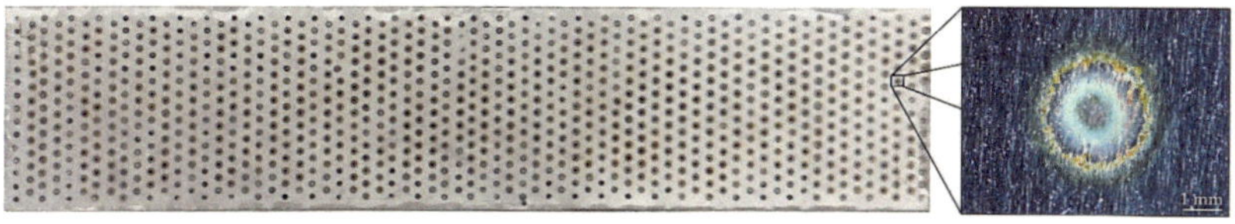

Figure 11. Example of the dark marks on the metal surface of the small-sized FMLP (MPin-C-MPin hexagonal 5 mm with dimensions ≈350 × 80 × 4 mm) caused by pin deposition process.

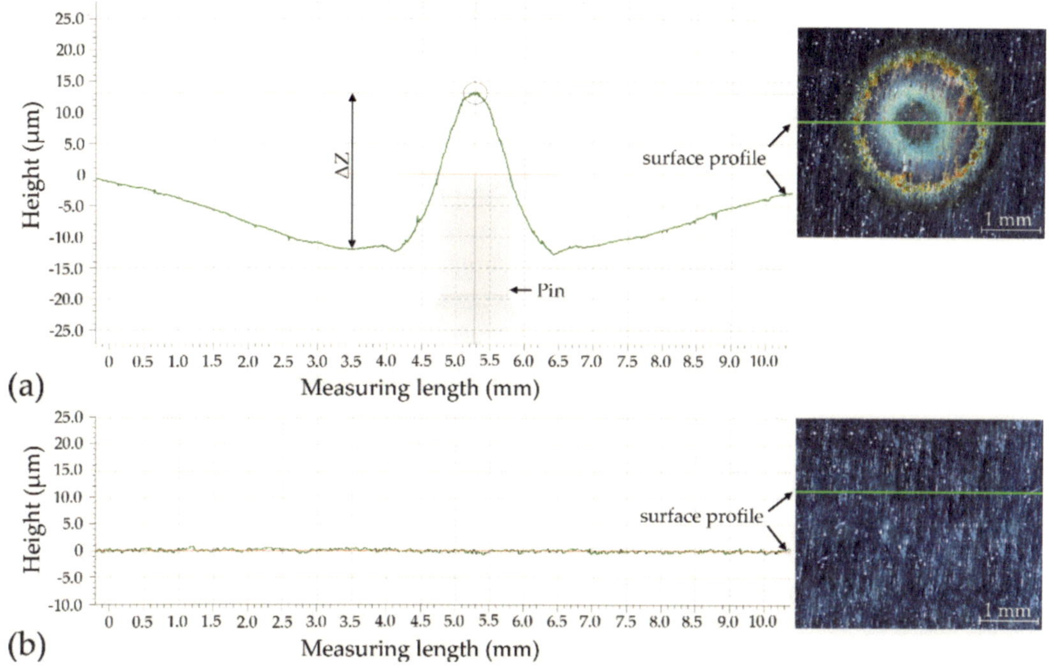

Figure 12. Typical surface profiles of the small-sized panels, where: (a) with pins (showing sharp waviness); (b) without pins (without waviness).

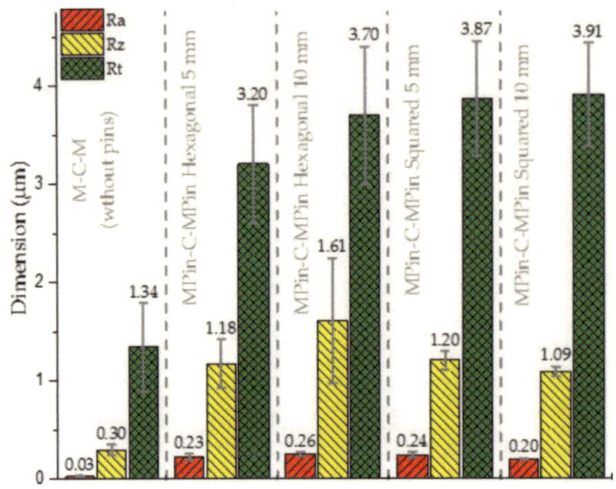

Figure 13. Results of Ra (mean), Rz (maximum) and Rt (total) obtained on small-sized FMLPs surfaces.

of the absolute values of the roughness peaks in relation to a midline within the measurement length. Rz is defined as the highest value of the roughness (peaks), that is, the roughness measured in segments defined by the cut-off, which is presented in the measuring path. Finally, Rt corresponds to the vertical distance between the highest peak and the deepest valley in the measurement length, obtained within the segments defined by the cut-off. These roughness representations were chosen based on the studies of De Chiffre et al. [22] and De Chiffre [23], which demonstrated the parameters most used in industry. The results show that all types of panels have roughly the same roughness values of Ra, Rz and Rt (considering the mean standard deviation). However, the deposition of the pins promoted an increase of roughness in the regions of the marks, in relation to the conventional panel.

Thus, it was concluded that the deposition of the pins by CMT PIN process, through its thermo-mechanical principle (electric arc with advancement and retraction of the wire), changes the surface profile of a metallic sheet (0.4 mm thickness), but only on a microscopic scale (at slightly less than 30 μm in terms of waviness and at just over 0.20 μm in terms of roughness). In this way, the elimination of staining due to pin deposition already largely removes the cosmetic drawbacks.

4. Destructive evaluation

In order to evaluate the influence of the pins as anchorages between metal and composite inside FMLPs, mechanical tests were employed until failure. Drop-weight test (low-speed impact test) was performed to assess the capacity of the FMLPs to dissipate impact energy. Buckling test, after impact (drop-weight test), was used to evaluate the performance of the FMLPs in such loading condition. And shear test (Iosipescu) was carried out to survey the effects that the pins have on the delamination of FMLPs.

4.1. Drop-weight testing

The FMLPs reinforced with pins, as conceived in this work, and the reference panels [M-C-M (without pins)] were submitted to impact damage by drop-weight testing. The small-sized FMLPs produced were cut in half their length, resulting in specimens of ≈175 × 80 × 4 mm each, and thus two specimens were used for each FMLP type. The aim with this test was to verify whether the pins would have positive or deleterious effects on the FMLPs concerning their capacity to absorb impact energy. Based on the ASTM D7136 standard [24], a rig to impose free fall (from around 1850 mm of height) of a constant mass (2.326 kg) over the small-sized panel surface was devised (**Figure 14(a)**). This mass was composed of a 28.5 mm spherical head made of hard steel attached to a plain carbon steel cylinder (50 mm of diameter and 150 mm of length). The rig included a latching device for ensuring no mass bouncing (single impact). A commercial high-speed camera filming at 2000 frames per second with 90 mm f/2.5 macro lens and frontal lighting was employed to quantify the energy (based on mass speed) involved in the impacts. The free fall height aimed a potential energy sufficient for causing apparent damage at impact, which resulted in 10.5 J per each millimeter of panel thickness (gravitational acceleration considered as 9.81 m/s^2). **Figure 14(b)** shows the upper and lower surfaces of all types of panels after impact.

High-speed images were used for determination of the falling/raising mass velocities immediately before/after impact, based on displacements (visualized from frame to frame) of its spherical head lower surface and respective time lapses, as seen in **Figure 15(a)**. A fitting curve, taking into account the non-uniform rectilinear motion due to gravity of the falling/raising mass, was figured out for each panel (including replications) and the velocities at the panel upper surface level were estimated by extrapolation. The velocity right at the end of the fall (actual impact) is referred as impact velocity and the velocity right at the beginning of the rebound as return velocity, which resultant average levels varied respectively from 5.81 to 5.96 m/s, as indicated in **Figure 15(b)**. According to Ursenbach et al. [25], the drop-weight test applied was classified as of low impact velocity (between 1 and 10 m/s). Farooq and Myler [26] consider an impact as of low velocity when an object impacts a target without penetrating it, situation observed for all panels tested in this present work.

The velocities involved in the impacts, in turn, were used to calculate the impact and return energies. Impact energy was considered as the kinetic energy of the falling mass just before actual impact (fall height tending to zero–panel upper surface level before impact). Analogously, return energy was taken as the kinetic energy of the raising mass at the beginning of the rebound after the impact (rebound height tending to zero–panel upper surface level before impact). Energy dissipation during impact was assumed as the relative drop in the kinetic energy due to impact. The impact energy quantities were represented by two ways: energy and specific energy (considering the panel mass density), as presented in **Figure 16**. As seen, the impact energy was always around 40 J, being the small fluctuation probably due experimental errors. The return energies and energy dissipations were also similar for all FMLPs types. In general, the presence of pins as anchorages inside the FMLPs did not seem to have any significant effect concerning the capacity of the panels to absorb impact energy. Therefore, the pins, at least for the impact conditions applied, did not make the FMLPs more brittle. In addition, the change in the deposition pattern of the pins, at least for the remaining conditions, did not show any effect concerning the capacity of the panels to dissipate impact energy.

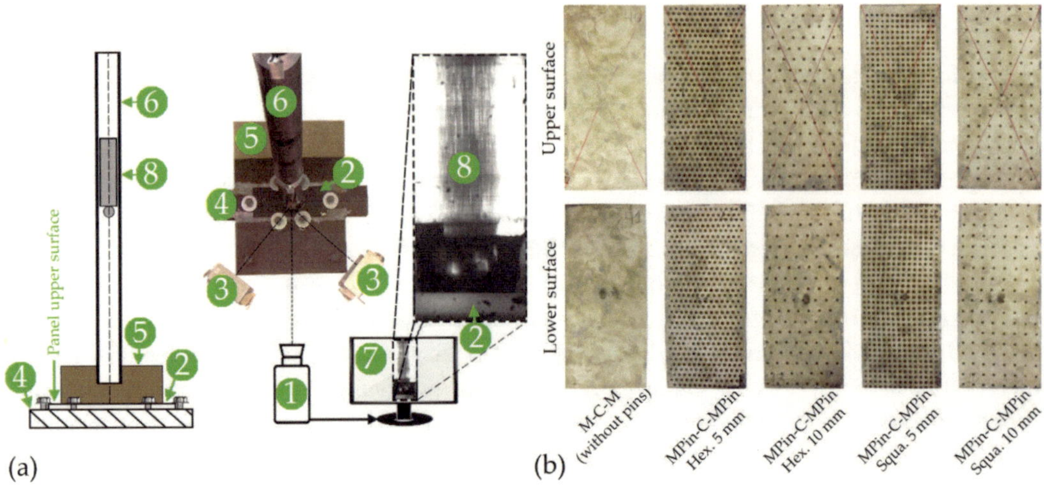

Figure 14. (a) Schematic frontal and top views of the drop-weight test rig, where: 1—high-speed camera with lens; 2—small-sized panel; 3—halogen lamp of 1 kW (2 units); 4—high-inertia base with central clearance hole (125 mm long, 75 mm wide and 25 mm deep); 5—flat background, perpendicular to the base; 6—mass guidance tube; 7—PC for image viewing; 8—mass with spherical head; (b) upper and lower surfaces of all types of panels (≈175 × 80 × 4 mm) after impact.

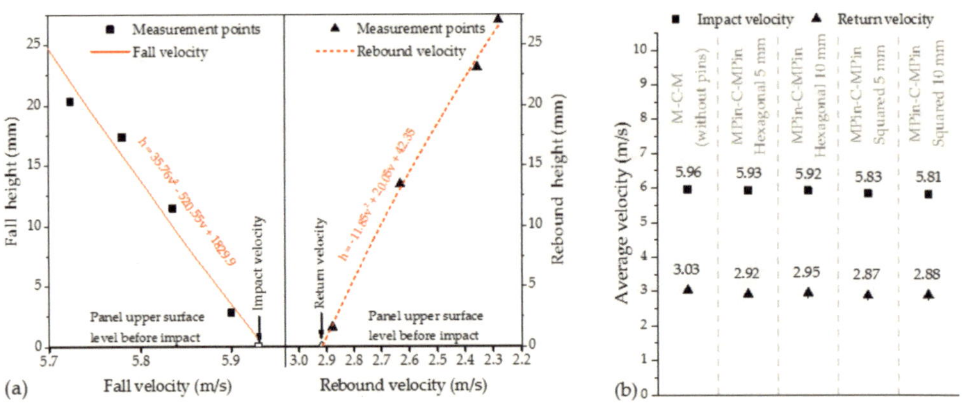

Figure 15. (a) Extrapolation of the falling/raising mass velocity at impact and return (MPin-C-MPin hexagonal 5 mm panel, as example); (b) average impact and return velocities for each panel (h = height; v = velocity).

4.1.1. Damage characterization (damage depth profile) after impact

The through-thickness damage extent, i.e., the depth profile of the impact damage (permanent deformation), was found by measuring the vertical displacement of the central transversal and central longitudinal lines drawn in all panels before drop-weight testing. The edges of the panels, both in length and width, were taken as references without permanent deformation after impact. A commercial manual-floating-type coordinate measuring machine, with 1 μm of resolution, was used for taking the measurements. The damage profile was determined by scanning the upper and lower surfaces of the panels, as exemplified in **Figure 17**. As seen, the measuring mesh was reduced to 5 mm near the damage area, against 10 mm in the rest of the surface of the panels.

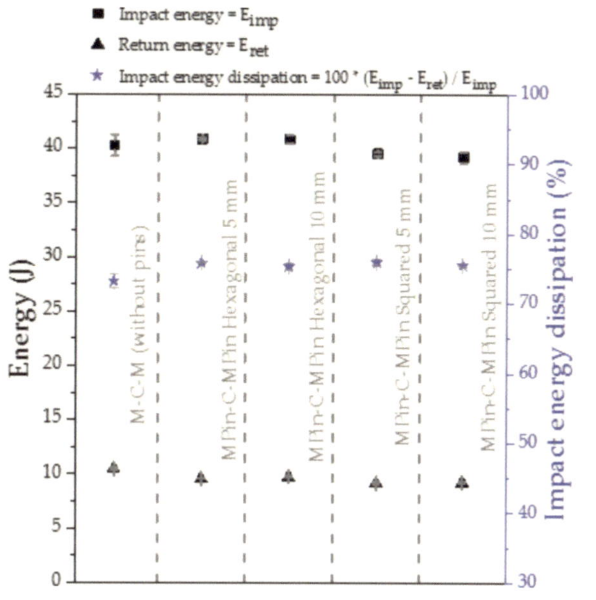

Figure 16. Average impact and return energies and consequent energy dissipation during drop-weight testing.

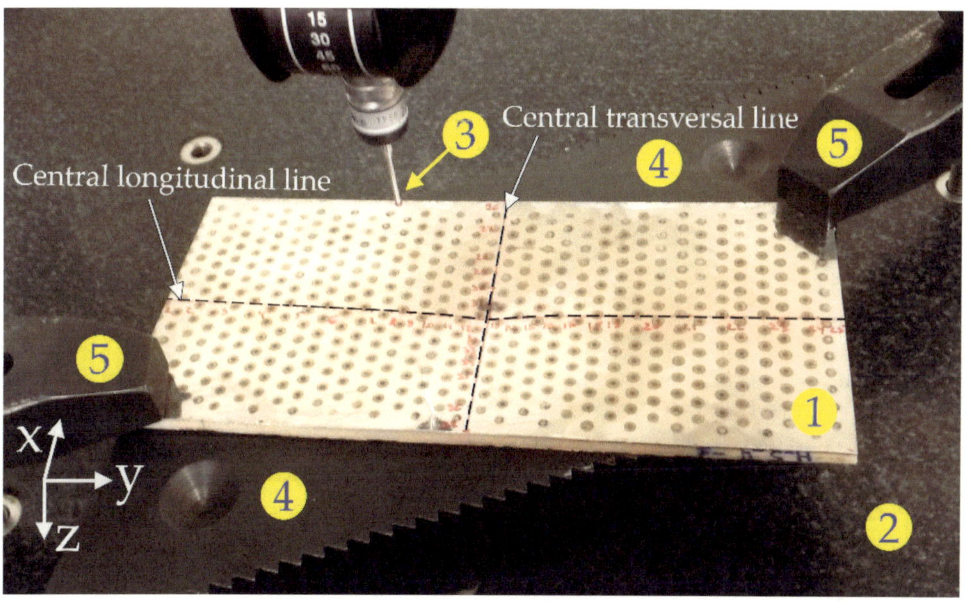

Figure 17. General setup for damage depth profile determination (upper panel surface, as example), where: 1—damaged small-sized panel (MPin-C-MPin hexagonal 5 mm panel, as example); 2—work table; 3—3D digital probe (touching head); 4—support; 5—clamping.

The average damage depth profiles of each panel are shown in **Figure 18(a)**. As the profiles were nearly longitudinally and transversally symmetrical, only half of the panel's length and width are represented. In general, the damage profiles found in each surface of the FMLPs with pins were similar, longitudinally as well as transversally, to the profiles found in the conventional

FMLP [M-C-M (without pins)]. However, the presence of pins tends to change the permanent deformation of the panels after impact as complementally shown in **Figure 18(b)**. It is also noted that the upper and lower surfaces exhibited different typical deformation profiles. In all panel types, the depth profile was more concentrated (shorter and shallower) in the upper surface than in the lower surface. The falling mass head might have tended to replicate its spherical contour on the upper surface, with the extent limited by the panel own thickness, which would work as restriction against deformation. The final damage extent at the lower surface, in contrast, is limited by the edges of the central clearance hole of the base of the drop-weight test rig, which are reasonably far away. It is even possible to observe that the change in damage depth profile is more gradual along the central longitudinal line (edges of support 125 mm apart) and more abrupt along the central transversal line (edges of support 75 mm apart). As the conventional FMLPs [M-C-M (without pins)] and FMLPs with pins had equal number of prepreg layers, they behaved similarly, consequently dissipating a close amount of energy during impact (**Figure 16**).

For a better understanding of the results of the least permanent deformed FMLPs, MPin-C-MPin squared 5 mm and MPin-C-MPin hexagonal 10 mm (**Figure 18(b)**), they were analyzed visually. In **Figure 19**, it is possible to notice that, in the case of these FMLPs, the spherical

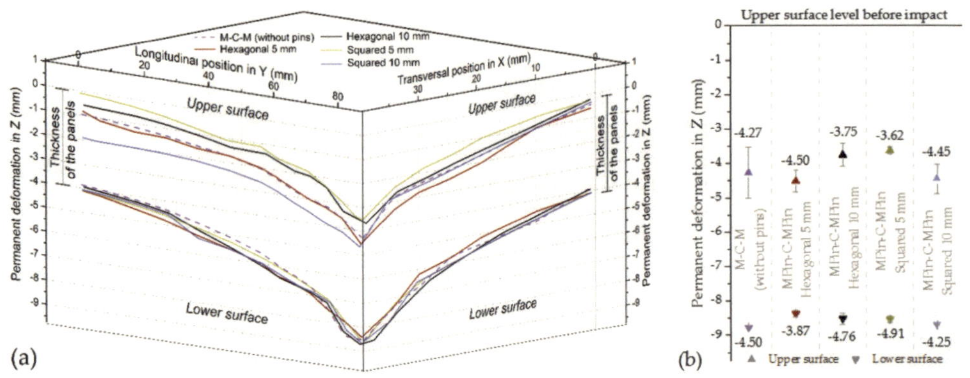

Figure 18. (a) Depth profiles of the damages produced in each panel (average of 2 samples) from 3D digital probing; (b) average maximum permanent deformation in each panel.

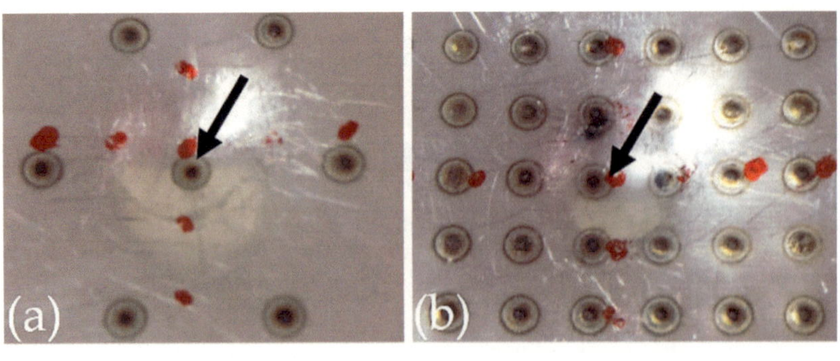

Figure 19. Zoom of the impact region of the upper surfaces of specific FMLPs after the drop-weight test, where: (a) MPin-C-MPin hexagonal 10 mm; (b) MPin-C-MPin squared 5 mm (the arrow indicates the deepest location of the panel after impact deformation).

head of the impactor reached the surface region with the deposited pin. In turn, this caused less permanent deformation in the panels with pins in comparison to the others. This indicates the pins acted holding the metal sheet to the composite, restraining delamination and external deformation.

4.2. Buckling after impact

According to Ishikawa et al. [27] and De Freitas and Reis [28], the resistance of composite panels to in-plane compressive stresses is strongly impaired by the presence of delamination-type damages, culminating in expressive reductions in buckling resistance of components. In that way, all types of FMLPs were subjected to buckling test after impact (drop-weight test) as investigation on the influence of pins on damage tolerance. For each type of small-sized FMLP two samples were tested for buckling. An electromechanical universal testing machine was used. Upper and lower supports for the panels were designed and built to allow proper alignment and fixture for the test, as shown in **Figure 20**. In accordance with ASTM D7137 standard [29], the testing speed (upper head moving rate) was always set at 1.25 mm/min.

Figure 21(a) shows the general evolution of axial displacement *versus* load in the buckling tests of all types of FMLPs after impact damage with compression, buckling and failure phases as detailed by Skhabovskyi et al. [10]. The mean values resulting from two tests of maximum compressive load, failure load, axial displacements at maximum load and failure are shown in **Figure 21(b)**. The axial displacement of the failure is shown as an alternative correlated with the lateral deflection of the panels tested. As seen, pinned FMLPs tolerated a higher maximum load, having close performances with each other, except the MPin-C-MPin squared 10 mm panel, which was able to withstand a smaller maximum load but still larger compared to the conventional panels [M-C-M (without pins)].

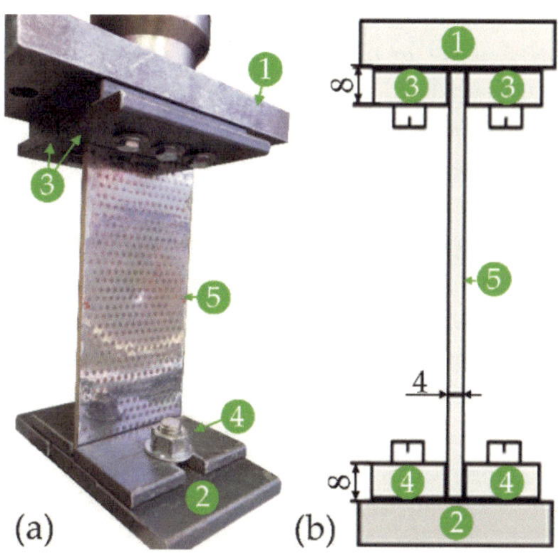

Figure 20. (a) Image of small-sized FMLP (MPin-C-MPin hexagonal 5 mm, as example) assembled in the testing machine; (b) schematic side view of the buckling set up, where: 1—moving upper head; 2—fixed lower head; 3—upper support (two parts); 4—lower support (two parts); 5—panel; 6—support fixture side plates; 7—auxiliary holding bar; 8—auxiliary clamps (dimensions in mm).

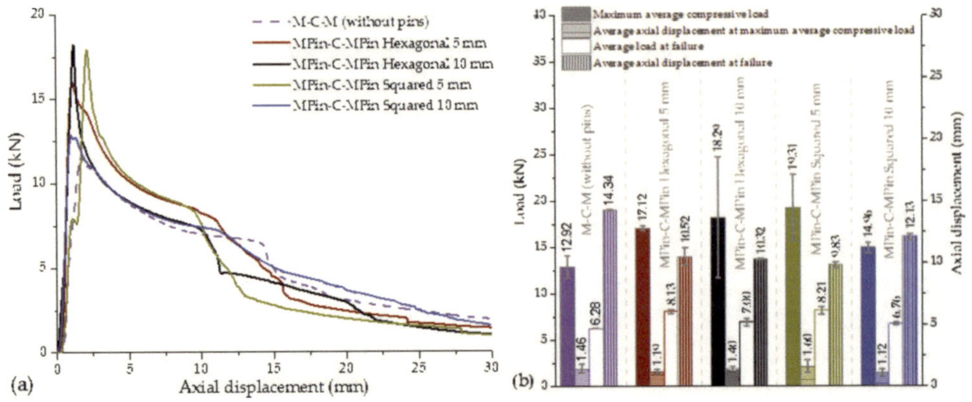

Figure 21. (a) Typical evolution of axial displacement *versus* load in buckling tests of all types of panels after impact damaging; (b) maximum average compressive load, average load at failure and correspondent axial displacements after buckling tests of all types of panels after impact damaging.

All types of the FMLPs achieved loads of maximum compression around 1.35 mm of axial displacement. **Figure 21(a)** shows that FMLPs with higher pin density (5 mm spacing) were able to withstand higher loads during deflection (buckling phase), followed by pinned panels in lower spacing patterns (panels MPin-C-MPin squared 10 mm and MPin-C-MPin hexagonal 10 mm in this order). As shown in **Figure 21(a)**, conventional FMLPs (without pins) reached a high load value at failure, still with a high axial displacement value at this moment, compared to FMLPs with pins. However, conventional FMLPs (without pins), after reaching a high value of failed load, showed a rapid unloading after the rupture (catastrophic failure).

MPin-C-MPin squared 5 mm and MPin-C-MPin hexagonal 10 mm panels supported higher maximum compression and failure loads (**Figure 21(b)**), probably because of their lower maximum permanent deformation (**Figure 18(b)**). It was also noted that MPin-C-MPin squared 5 mm and MPin-C-MPin hexagonal 10 mm panels did not show a good repeatability of their high mean maximum compressive load, as evidenced by a high average standard deviation in **Figure 21(b)**. This probably happened because the spherical head of the impactor hit the pin region in one of two tests performed, thus leading to the smallest deformation. The other FMLPs showed similar permanent deformation, which is supported by good repeatability of the loads and displacements in the buckling test. In this case, the pins possibly tend to retard the propagation of debonding between metal sheets and composite, for anchoring them to each other, and even tend to retard the spread of delamination in the composite, for acting as clamps that hold most of its layers together between the metal sheets and pins ball-heads. That is, the pins tend to delay the propagation of the delamination between the metal sheets and the composite by their anchoring and even tend to delay the propagation of the delamination in the composite, acting as staples that support most of their layers between the metal sheets and ball-head pins.

Figure 22 shows the images of all types of panels with impact damage after the buckling test. As expected, all types of panels had their lateral deflection going from the upper surface (side of impact) towards the lower surface. All panels collapsed and ended folded with presence of corrugations transversally crossing the areas of damage. Still all FMLPs had the lower metal sheet

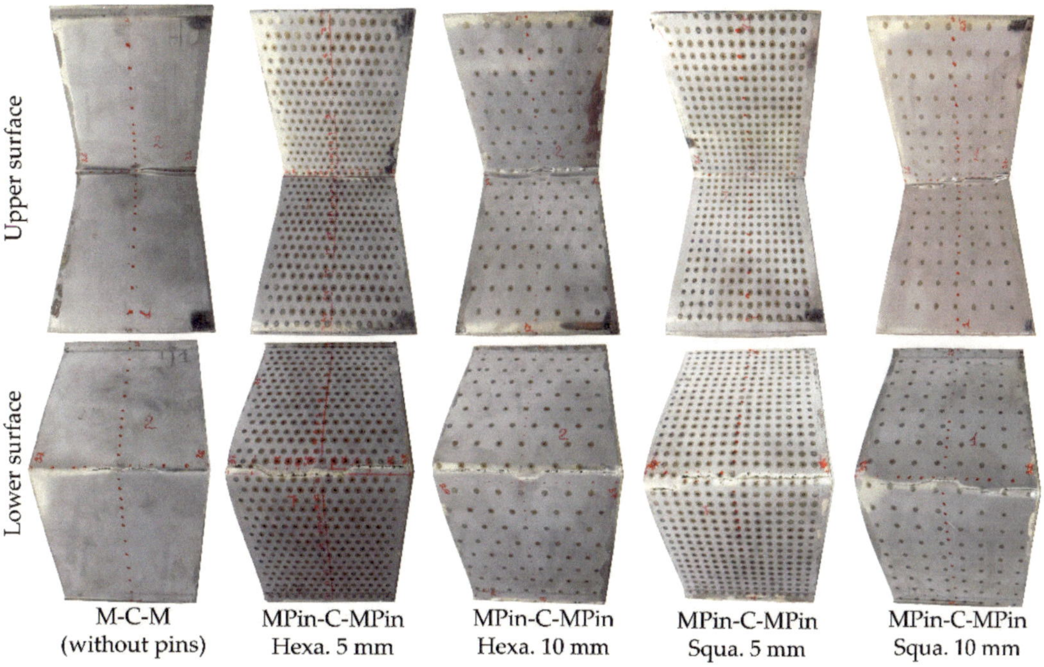

Figure 22. Typical appearance of all types of FMLPs with impact damage after buckling test.

ruptured transversely in the middle after the buckling test, including the conventional panel (without pins). But in the case of M-C-M (without pins), the fracture did not cross the entire width of the panel, concentrating closer to the damage region (**Figure 22**). It was also noticed that the ruptures of the metallic sheets happened between the metallic pins (spacings), because the pins acted as anchoring of the metallic sheets and because the tensile tensions were present in the inferior metallic sheets (side opposite to the impact) between the pins. It was noted that the M-C-M type panels (without pins) with impact damage ended the buckling test with the metal sheets slightly separated from the composite. In contrast, the FMLPs with pins exhibited metal sheets-composite debonding notably concentrated around the damage areas. This fact ratifies that the pins effectively anchor the metal sheets to the composite, even after impact damage.

4.3. Iosipescu shear test

To evaluate the influence of the deposition of the pins in the FMLP on shear and delamination resistance, a shear strength test was used, as proposed for the first time in 1967 by Iosipescu. According to Le Bourlegat [30], the Iosipescu shear test uses a simple test body because it is flat and achieves a pure and uniform shear stress-strain state in the region of the notch imposed on the specimen. **Figure 23(a)** shows a simplified view of this test performed on the universal machine for mechanical testing. **Figure 23(b)** shows the dimensions of the specimen used and defined by ASTM D5379 standard [31]. In addition, every Iosipescu shear test procedure was based on this standard, keeping the movable jaw displacement speed equal to 0.5 mm/min. The test was carried out on five specimens taken from each of the five

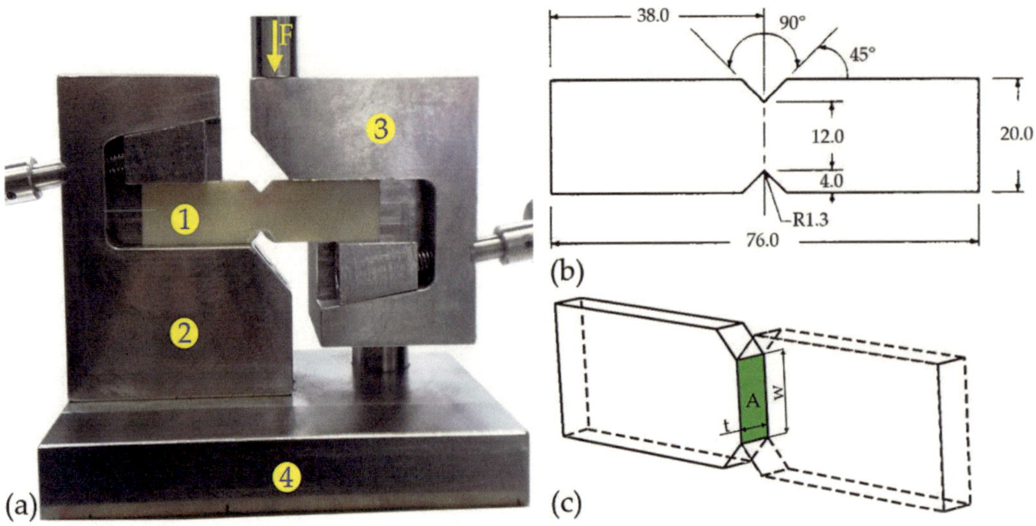

Figure 23. (a) Simplified view of the Iosipescu shear test, where: 1—specimen; 2—fixed jaw; 3—mobile jaw; 4—base; F—applied load; (b) dimensions (in mm) of the specimens used in the Iosipescu shear test (adapted from ASTM D5379 standard [31]); (c) area of the specimen used for shear stress calculations (adapted from Le Bourlegat [30]).

types of small-sized FMLPs. **Figure 23(c)** shows the area (A) of the plane between the notches of the specimen (w ≈ 12 mm and t ≈ 4 mm) along which the load (F) was applied. To calculate the shear stress (τ), Eq. (1) was used.

$$\tau = \frac{F}{w * t} \quad (1)$$

Figure 24 presents the typical appearance of FMLPs specimens before and after the Iosipescu shear test. **Figure 25** shows typical displacement *versus* load curves, respectively, for each type of panel in the Iosipescu shear test. During the tests a good repeatability of the results was observed for each of the five types of FMLPs specimens. **Figure 26** shows column diagrams with average values of maximum load, maximum load displacement, and maximum shear stress. It is observed that the specimens of the panels with the largest number of pins (5 mm spacing), regardless of their deposition pattern, showed the highest average maximum load (≈6.61 kN). The panels with lower pin density (10 mm spacing) exhibited a maximum load a bit lower (≈5.83 kN), regardless of the deposition pattern. Finally, the conventional FMLP specimens [M-C-M (without pins)] showed the lowest maximum load (≈5.04 kN), approximately 25% below the results with the specimens of the panels with spacing of 5 mm.

The specimens of FMLPs with pins reached a maximum load displacement of approximately 2.91 mm (**Figure 26**), except for the specimens of the MPin-C-MPin squared 10 mm panel that showed a lower displacement value at the moment of failure (1.44 ± 0.06 mm). This probably happened because of the pin position inside both notches (**Figure 24**). As there was a pin very close to the edge of the notch, there may have been stress concentration right at this point, thus breaking the specimens of this case in advance and leaving no room for further displacements.

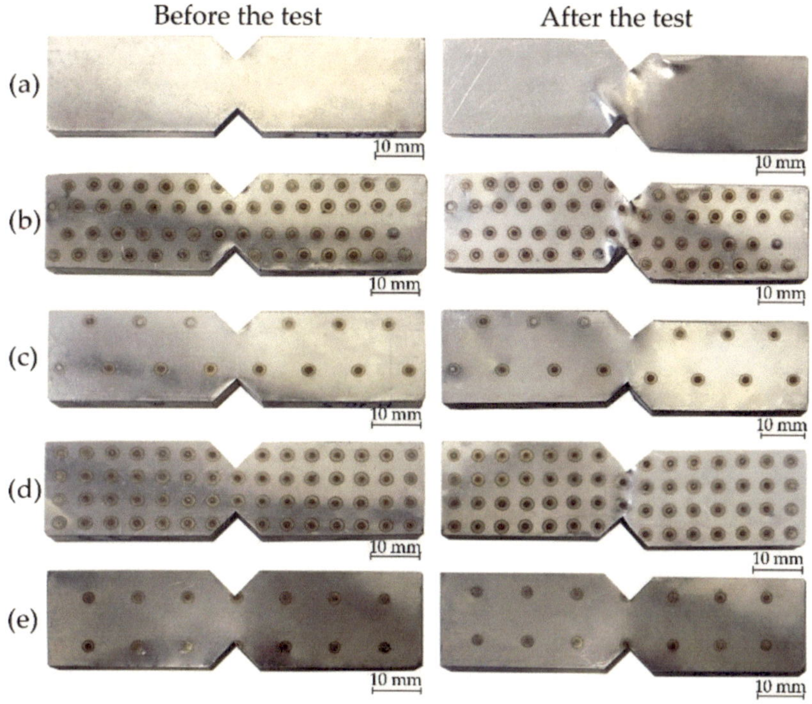

Figure 24. Typical aspect of the five types of FMLPs specimens before (left) and after (right) the Iosipescu shear test, where: (a) M-C-M (without pins); (b) MPin-C-MPin hexagonal 5; (c) MPin-C-MPin hexagonal 10; (d) MPin-C-MPin squared 5; (e) MPin-C-MPin squared 10.

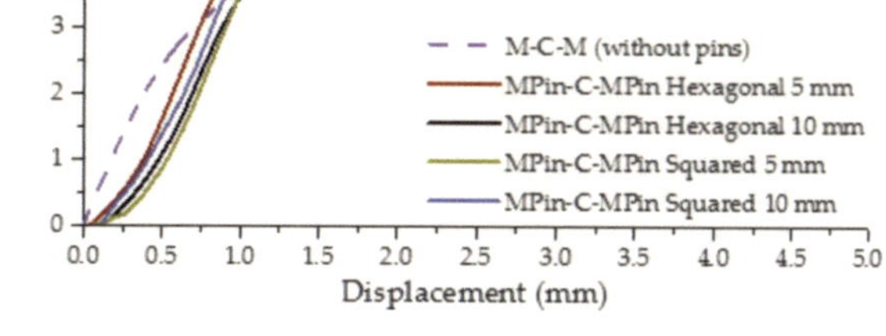

Figure 25. Typical displacement *versus* load curves for each FMLP type in the Iosipescu shear test.

In general, the panels with pins showed better results of maximum load, demanding a higher value of shear stress to break. Probably, this result was caused because the contact between the metallic sheets and the composite of the pinned specimens are more anchored due to the presence of the pins. In such a way, the greater number of pins (5 mm spacing) provided

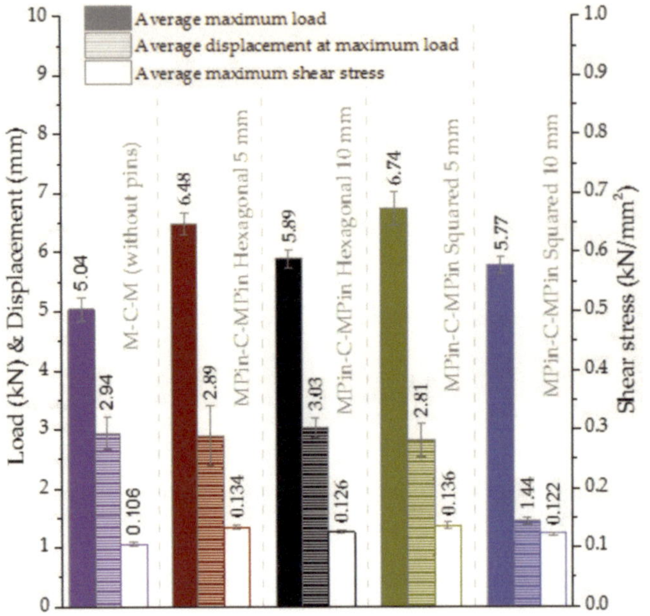

Figure 26. Average results obtained through five measurements of maximum load, displacement at maximum load and maximum shear stress of the Iosipescu shear test.

a better union, consequently showing better results. Possibly, failures in pinned specimens would be less catastrophic after reaching the maximum load (after failure), which should be exploited in future work.

5. Conclusions

Considering the conditions applied in the evaluation of pins deposited by CMT PIN process in FMLPs, the following conclusions are listed according to the type of testing.

Non-destructive evaluation:

- By means of the four natural frequency modes obtained after the modal analysis, the presence of the pins in the FMLPs did not result in a significant stiffness increase. There are, in this case, probably two factors acting in opposite directions; the pins would lead to an increase in stiffness due to the anchoring action between metal and composite, but on the other hand the mass increase with more pins inside the panels is a stiffness-reducing factor as in any structure. However, the presence of the pins in the FMLPs significantly increased the damping factor, dissipating the applied vibration wave. Pin deposition was able to reduce the propagation of vibrations at low frequencies (from 0 to 800 Hz) by up to about five times (for some resonance modes).

- CMT PIN process, through its thermo-mechanical working principle, changes the surface profile of the metal sheets on the opposite face in the region of deposition, but only on a microscopic scale. In addition, thermoxidation occurs in these regions, which is not a problem because this inconvenience could be avoided/minimized by the application of purge gas (supplementary protection to that used near the electric arc and inert).

Destructive evaluation:

- In terms of impact energy dissipation, all FMLPs with pins exhibited similar performance, generally equivalent to the conventional FMLP (without pins). Besides, the addition of weight by the pins as anchorages does not penalize the capacity of the panels to dissipate impact energy, as all panel types dissipated similar levels of specific energy during impact. Therefore, the pins did not make the FMLPs more brittle and the change in their deposition pattern did not show any significant effect concerning the capacity of the panels to dissipate impact energy.
- Concerning the damage depth profiles caused by drop-weight testing, all FMLPs with pins suffered damages similar to that found in the conventional FMLP (without pins).
- Regarding damage tolerance, the FMLPs with pins exhibited a less catastrophic trend, i.e., achieving significantly higher loads at longer axial displacements in buckling test after impact and the pins tend to retard the debonding propagation between metal sheets and composite, for anchoring them to each other. The pins also hold back the delamination spreading in the composite, for clamping the layers together between the metal sheets and ball-heads of the pins.
- The anchoring effect the pins have on the FMLPs was confirmed through the Iosipescu shear test. Generally, the panels with pins exhibited higher shearing loads in relation to the conventional panel (without pins). The FMLPs with the highest number of pins (5 mm spacing), regardless of the deposition pattern, presented the highest maximum loads and displacements at the moment of failure.

Acknowledgements

The authors acknowledge the Center for Research and Development of Welding Processes (Laprosolda) and Laboratory of Structural Mechanics Prof. José Eduardo Tannús Reis, both at UFU, and the Department of Materials and Technology, at UNESP-Guaratinguetá, for providing laboratorial infrastructure and support, and ALLTEC Materiais Compostos, for providing composite materials. They also acknowledge the Brazilian agencies CNPq (Procs. 302863/2016-8 and 303224/2016-9), CAPES (PROEX 0309/2015), and FAPEMIG (Proc. 11304), for research grants and scholarships.

Author details

Ruham Pablo Reis[1]*, Iaroslav Skhabovskyi[1], Alberto Lima Santos[2], Leonardo Sanches[1], Edson Cocchieri Botelho[2] and Américo Scotti[1,3]

*Address all correspondence to: ruhamreis@mecanica.ufu.br

1 Centro Para Pesquisa e Desenvolvimento de Processos de Soldagem—LAPROSOLDA, Universidade Federal de Uberlândia—UFU, Uberlândia, Brazil

2 Departamento de Tecnologias e Materiais, Universidade Estadual Paulista—UNESP, Guaratinguetá, Brazil

3 Department of Engineering Science, Production Technology West, Division of Welding Technology, University West, Trollhättan, Sweden

References

[1] Cortes P, Cantwell WJ. The prediction of tensile failure in titanium-based thermoplastic fibre-metal laminates. Composites Science and Technology. 2006;**66**:2306-2316. DOI: 10.1016/j.compscitech.2005.11.031

[2] Sinmazçelik T, Avcu E, Bora MÖ, Çoban O. A review: Fibre metal laminates, background, bonding types and applied test methods. Materials and Design. 2006;**32**:3671-3685. DOI: 10.1016/j.matdes.2011.03.011

[3] Salve A, Kulkarni R, Mache A. A review: Fiber metal laminates (FML's)—Manufacturing, test methods and numerical modeling. International Journal of Engineering Technology and Sciences (IJETS). 2016;**6**:71-84. DOI: 10.15282/ijets.6.2016.10.2.1060

[4] Seydel R, Chang F. Impact identification of stiffed composite panels: I. System development. Smart Materials and Structures. 2001;**10**:354-369. PII: S0964-1726(01)18317-5

[5] Vlot A. Impact loading on fibre metal laminates. International Journal of Impact Engineering. 1996;**18**:291-307. DOI: 10.1016/0734-743X(96)89050-6

[6] Jahn J, Weeber M, Boehner J, Steinhilper R. Assessment strategies for composite-metal joining technologies—A review. Procedia CIRP. 2016;**50**:689-694. DOI: 10.1016/j.procir.2016.05.034

[7] Fronius. Current Welding Practice: CMT Technology. 1st ed. Düsseldorf: DVS Media GmbH; 2014. 96 p. ISBN: 978-3-945023-36-5

[8] Ucsnik S, Scheerer M, Zaremba S, Pahr HD. Experimental investigation of a novel hybrid metal-composite joining technology. Composites Part A: Applied Science and Manufacturing. 2010;**41**:369-374. DOI: 10.1016/j.compositesa.2009.11.003

[9] Graham DP, Rezai A, Baker D, Smith PA, Watts JF. The development and scalability of a high strength, damage tolerant, hybrid joining scheme for composite-metal structures. Composites Part A: Applied Science and Manufacturing. 2014;**64**:11-24. DOI: 10.1016/j.compositesa.2014.04.018

[10] Skhabovskyi I, Batista NL, Damato CA, Reis RP, Botelho EC, Scotti A. Appraisal of fiber-metal laminate panels reinforced with metal pins deposited by CMT welding. Composite Structures. 2017;**180**:263-275. DOI: 10.1016/j.compstruct.2017.07.043

[11] Leme LMM. Estudo comparativo de técnicas de análise modal experimental no domínio da frequência em uma barra na condição de contorno livre-livre [trabalho de conclusão de curso]. Pato Branco: Universidade Tecnológico Federal do Paraná; 2014. Available from: http://repositorio.roca.utfpr.edu.br/jspui/bitstream/1/6094/1/PB_DAMEC_2014_2_03.pdf Accessed: Apr 23, 2017

[12] Rao SS. Vibrações Mecânicas. 4th ed. São Paulo: Pearson Prentice Hall; 2008. 424 p. ISBN: 978-85-7605-200-5

[13] Bolina CC, Palechor EUL, Vásquez MPR, Nicacio WG, Gutierres, MPD, Lopez AAO. Vibrações: as frequências naturais estimada e experimental de uma estrutura. In: Congresso Nacional de Matemática Aplicada à Indústria (CNMAI); Caldas Novas, Brasil; Nov 18-21, 2014. p.1-5. Blucher Ed. Available from: http://pdf.blucher.com.br.s3-sa-east-1.amazonaws.com/mathematicalproceedings/cnmai2014/0038.pdf [Accessed: Apr 23, 2018]

[14] Lundkvist A. Modal Analysis Project [Internet]. 2010. Available from: http://www.student.ltu.se/~nadlun-5/pdfs/l7005a_report_main.pdf [Accessed: Apr 23, 2018]

[15] Soeiro NS. Análise Modal Experimental [Internet]. 2001. Available from: http://www.ebah.com.br/content/ABAAAgT_QAK/analise-modal-experimental [Accessed: Apr 23, 2018]

[16] Richardson MH, Formenti DL. Parameter estimation from frequency response measurements using rational fraction polynomials. In: 1st Proceedings of the International Modal Analysis (IMAC) Conference; November 1982; Orlando. USA. 1982. pp. 1-15. Available from: http://citeseerx.ist.psu.edu/viewdoc/download?doi=10.1.1.458.3592&rep=rep1&type=pdf [Accessed: Apr 23, 2018]

[17] Laliberte JF, Poon C, Straznicky PV, Fahr A. Applications of fiber-metal laminates. Polymer Composites. 2000;**21**:558-567. DOI: 10.1002/pc.10211

[18] Marsh G. Composites lift off in primary aerostructures. Reinforced Plastics. 2004;**48**: 22-27. DOI: 10.1016/S0034-3617(04)00193-6

[19] Botelho EC, Silva RA, Pardini LC, Rezende MC. A review on the development and properties of continuous fiber/epoxy/aluminum hybrid composites for aircraft structures. Materials Research. 2006;**9**(3):247-256. DOI: 10.1590/S1516-14392006000300002

[20] Associação Brasileira de Normas Técnicas. ABNT NBR NM ISO 1. ABNT Ed. Temperatura padrão de referência para medições industriais de comprimento. Brasília, DF: 1997

[21] Associação Brasileira de Normas Técnicas. ABNT NBR ISO 4288. ABNT Ed. Especificações geométricas do produto (GPS)—Rugosidade: Método do perfil—Regras e procedimentos para avaliação de rugosidade. Brasília, DF: 2008

[22] De Chiffre L, Lonardo P, Trumpold H, Lucca DA, Goch G, Brown CA, Raja J, Hansen HN. Quantitative characterization of surface texture. CIRP Annals—Manufacturing Technology. 2000;**49**(2):635-352. DOI: 10.1016/S0007-8506(07)63458-1

[23] De Chiffre L. Industrial survey on ISO surface texture parameters. CIRP Annals—Manufacturing Technology. 1999;**48**(3):74-77

[24] American Society for Testing and Materials. ASTM D7136/D7136M-15. Standard Test Method for Measuring the Damage Resistance of a Fiber-Reinforced Polymer Matrix Composite to a Drop-Weight Impact Event, West Conshohocken, PA. In: Annual Book of ASTM Standards. 2015

[25] Ursenbach DO. Penetration of CFP laminates by cylindrical indenters [thesis]. The University of British Columbia. Vancouver, Canada; 1995. Available from: https://open.library.ubc.ca/cIRcle/collections/ubctheses/831/items/1.0078475 [Accessed: Aug. 03, 2018]

[26] Farooq U, Myler P. Finite element simulation of damage and failure predictions of relatively thick carbon fibre-reinforced laminated composite panels subjected to flat and round noses low velocity drop-weight impact. Thin Walled Structures. 2016;**104**:82-105. DOI: 10.1016/j.tws.2016.03.011

[27] Ishikawa T, Sugimoto S, Matsushima M, Hayashi Y. Some experimental findings in compression-after-impact (CAI) tests of CF/PEEK (APC-2) and conventional CF/EPOXY flat plates. Composites Science and Technology. 1995;**55**:349-363. DOI: 10.1016/0266-3538(95)00079-8

[28] De Freitas M, Reis L. Damage growth analysis of low velocity impacted composite panels. Composite Structures. 1997;**38**(1-4):509-515. DOI: 10.1016/S0263-8223(97)00086-X

[29] American Society for Testing and Materials. ASTM D7137/D7137M-12. Standard Test Method for Compressive Residual Strength Properties of Damaged Polymer Matrix Composite Plates, West Conshohocken, PA. In: Annual Book of ASTM Standards. 2012

[30] Le Bourlegat LR. Processamento e Caracterização Híbrido Titânio/Fibra de Carbono/Epóxi [thesis]. São José dos Campos: Instituto Tecnológico de Aeronáutica (ITA); 2009

[31] American Society for Testing and Materials. ASTM D5379/D5379M-12. Standard Test Method for Shear Properties of Composite Materials by the V-Notched Beam Method, West Conshohocken, PA. In: Annual Book of ASTM Standards. 2012

Chapter 7

Design Optimization of Reinforced Ordinary and High-Strength Concrete Beams with Eurocode2 (EC-2)

Fedghouche Ferhat

Additional information is available at the end of the chapter

http://dx.doi.org/10.5772/intechopen.78734

Abstract

This chapter presents a method for minimizing separately the cost and weight of reinforced ordinary and high-strength concrete (HSC) T-beams at the limit state according to Eurocode2 (EC-2). The first objective function includes the costs of concrete, steel, and formwork, and the second objective function deals with the weight of the T-beam. All the constraints functions are set to meet the design requirements of Eurocode2 and current practices rules. The optimization process is developed through the use of the generalized reduced gradient (GRG) algorithm. Two example problems are considered in order to illustrate the applicability of the proposed design model and solution methodology. It is concluded that this approach is economically more effective compared to conventional design methods used by designers and engineers and can be extended to deal with other sections without major alterations.

Keywords: cost and weight minimization, reinforced ordinary and high-strength concrete beams, Eurocode2 (EC-2), nonlinear optimization, algorithm

1. Introduction

Structural elements with T-shaped sections are frequently used in industrial construction. They are used for repeated and large structures because they are cost effective when using the optimum cost design model which is of great value for designers and engineers. Compression reinforcement is not often required when designing the T-beams sections. One of the great advantages of T-beams sections is the economy in the amount of steel needed for reinforcement. The objective function is usually simplified to represent the weight, disregarding the costs of shaping and the construction details. However, the economy aspects in terms of costs and gain achieved should be the area where scope exists for extending the research works [1–4].

Recent developments in the technology of materials have led to the use of the high-strength concrete (HSC); this is mainly due to its efficiency and economy. The reduction in the quantities of construction materials has enabled both a gain in weight reduction and in the foundation's cost. HSC has a high compressive strength in the range of 55–90 MPa; it not only has the advantage of reducing member size and story height, but also the volume of concrete and the area of formwork. In terms of the amount of steel reinforcement, there is a substantial difference between the normal-strength concrete structures compared to high-strength concrete structures [5, 6]. In this chapter, not only does it presents the minimum weight design but it presents a detailed objective function that considers the ratio cost not the absolute cost with sensitivity analysis of this cost ratio as well. It considers both shaping and material costs. The generalized reduced gradient (GRG) method is used to solve nonlinear programming problems. It is a very reliable and robust algorithm; also, various numerical methods have been used in engineering optimization [7–12].

This work shows a method for minimizing separately the cost and weight of reinforced ordinary and high-strength concrete (HSC) T-beams at the limit state according to Eurocode2 (EC-2). The first objective function includes the costs of concrete, steel and formwork, whereas the second objective function represents the weight of the T-beam; all the constraints functions are set to meet the ultimate strength and serviceability requirements of Eurocode2 and current practices rules. The optimization process is developed through the use of the generalized reduced gradient algorithm. Two example problems are considered in order to illustrate the applicability of the proposed design model and solution methodology. It is concluded that this approach is economically more effective compared to conventional design methods applied by designers and engineers and can be extended to deal with other sections without major alterations.

2. Limit state design of reinforced concrete T-section under bending

In accordance with EC-2 [13], the assumptions used at the limit state for the typical reinforced T-beam-cross section are, respectively, illustrated in **Figure 1(a)–(c)**.

In the linear strain diagram of **Figure 1b**, the symbols ε_s and ε_{cu3} designate steel strain and the ultimate strain for the rectangular stress distribution compressive concrete design stress–strain relation. The parameter α represents the relative depth of the compressive concrete zone and the plastic neutral axis is located at the distance αd from the upper fiber for the ultimate limit state design, and x is the depth of elastic neutral axis for serviceability limit state design. In the assumed uniformly distributed stress diagram of **Figure 1c**, f_{cd} is the design value of concrete compressive strength, γ_c is the partial safety factor for concrete and f_{ck} is the characteristic compressive cylinder strength of ordinary or HSC at 28 days. In accordance with EC-2, the possibility of working with rectangular stress distribution is offered. This requires the

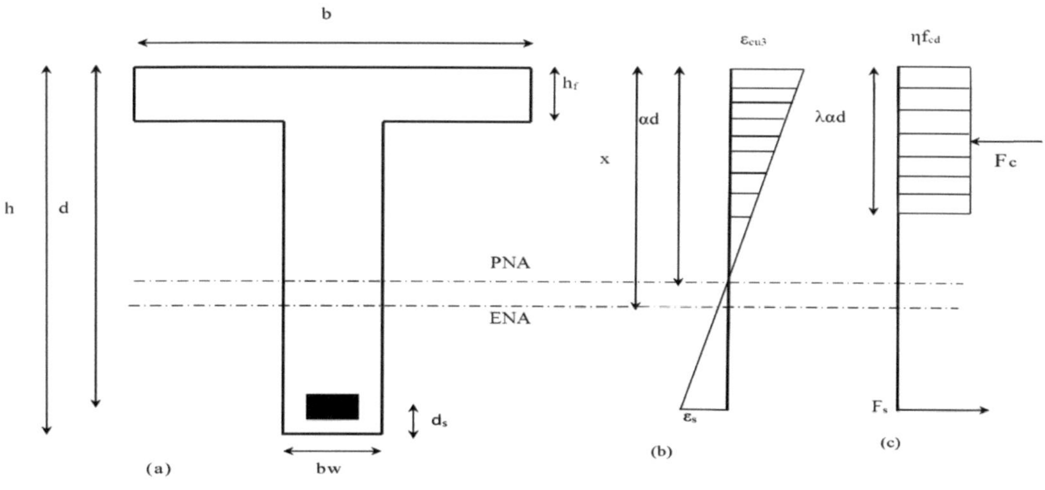

Figure 1. (a) Typical T-beam cross section; (b) strains at ultimate limit state and (c) stresses at ultimate limit state.

introduction of a factor λ for the depth of the compression zone and a factor η for the design strength. The λ and η factors are both linearly dependent on the characteristic strength f_{ck} in accordance with the following Equations [13]:

$$\lambda = 0.8 - \frac{f_{ck} - 50}{400} \tag{1}$$

$$\mu = 1.0 - \frac{f_{ck} - 50}{200} \tag{2}$$

with $50 \leq f_{ck} \leq 90$ MPa and $\lambda = 0.8, \eta = 1.0$ for $f_{ck} \leq 50$ MPa.

F_c and F_s denote the resultants of internal forces in the HSC section and reinforcing steel, respectively.

The design yield strength of steel reinforcement is $f_{yd} = f_{yk}/\gamma_s$ where f_{yk} is the characteristic elastic limit of steel and γ_s is the partial safety factor. In addition, the steel strain is considered unlimited in accordance with the Eurocode2 provisions. In this chapter, for an optimal use of steel, the strain must always be greater or equal to elastic limit strain, $\varepsilon_{yd} = f_{yd}/E_s$ where E_s represents the elasticity modulus for steel.

3. Formulation of the optimization problem

3.1. Design variables

The design variables selected for the optimization are presented in **Table 1**.

Design variables	Defined variables
b	Effective width of compressive flange
b_w	Web width
h	Total depth
d	Effective depth
h_f	Flange depth
A_s	Area of tension reinforcement
α	Relative depth of compressive concrete zone

Table 1. Definition of design variables.

3.2. Objective functions

3.2.1. *Cost function*

The objective function to be minimized in the optimization problems is the total cost of construction material per unit length of the beam. This function can be defined as:

$$C_0/L = C_c(b_w h + (b - b_w)h_f) + C_s A_s + C_f[b + 2h] \rightarrow \text{Minimum} \tag{3}$$

Thus, the cost function to be minimized can be written as follows:

$$C = \frac{C_O}{C_c L} = b_w h + (b - b_w)h_f + \left(\frac{C_s}{C_c}\right)A_s + \left(\frac{C_f}{C_c}\right)[b + 2h] \rightarrow \text{Minimum} \tag{4}$$

The values of the cost ratios C_s/C_c and C_f/C_c vary from one country to another and may eventually vary from one region to another for certain countries [14, 15].

3.2.2. *Weight function*

The weight function to be minimized can be written as follows:

$$W = (b_w h + (b - b_w)h_f)\rho \longrightarrow (\text{minimum}) \tag{5}$$

where

ρ is the density of the reinforced concrete T-beams and W is the unit weight per unit length of the reinforced concrete T- beams.

3.3. Design constraints

a. Behavior constraints:

$$M_{Ed} \leq \eta f_{cd}(b - b_w)h_f(d - 0,50h_f) + \eta\lambda f_{cd} b_w d^2 \alpha(1\text{–}0,5\lambda\alpha) \tag{6}$$

(external moment ≤ resisting moment of the cross-section)

$$\alpha = \left(\frac{f_{yd}}{f_{cd}}\right)\left(\frac{A_S}{\eta\lambda b_w d}\right) - \frac{(b - b_w)h_f}{\lambda b_w d} \tag{7}$$

(internal force equilibrium)

$$\frac{A_s}{b_w d} \geq p_{min} \tag{8}$$

(minimum steel percentage)

$$\frac{As}{b_w h + (b - b_w)h_f} \leq p_{max} \tag{9}$$

(maximum steel percentage)

In Eqs. (7) and (8) above, it is assumed that the neutral axis position is under the beam flange which ensures that the section behaves as the T-beam section shown in **Figure 1a**.

Conditions on strain compatibility in steel:

$$\varepsilon_{cu3}\left(\left(\frac{1}{\alpha}\right) - 1\right) \geq \frac{f_{yd}}{E_s} \tag{10}$$

(In the case of Pivot B, optimal use of steel requires that strains in steel must be limited to plastic region at the ultimate limit state (ULS).)

$$\lambda\alpha(1\text{–}0,5\lambda\alpha) \leq \mu_{limit} \tag{11}$$

(Compression reinforcement is not required.)

b. Shear strength constraint:

$$V_{Ed} \leq V_{Rd,max} = \nu_1 \frac{f_{cd} b_w z}{tg(\theta) + cotg(\theta)} \tag{12}$$

(external shear force ≤ resisting shear force)

c. Deflection constraint:

$$\frac{5wL^4}{384\, E_{cm} I_c} \leq \delta_{lim} \tag{13}$$

$$I_c = \frac{b_w h^3}{3} + \frac{(b - b_w)h^3}{3} + nA_s d^2 - A_h x^2 \tag{14}$$

$$A_h = b_w h + (b - b_w)h_f + nA_s \tag{15}$$

$$x = \frac{\frac{b_w h^2}{2} + \frac{(b-b_w)h_f^2}{2} + nA_s d}{A_h} \tag{16}$$

d. Geometric design variable constraints including rules of current practice:

$$h \geq \frac{L}{16} \tag{17}$$

$$\frac{d}{h} = 0.90 \tag{18}$$

$$0.20 \leq \frac{b_w}{d} \leq 0.50 \tag{19}$$

$$\frac{(b-b_w)}{2} \leq \frac{L}{10} \tag{20}$$

$$\frac{b}{h_f} \leq 8 \tag{21}$$

$$h_f \geq h_{fmin} \tag{22}$$

$$\frac{b}{b_w} \geq 3 \tag{23}$$

where:

μ_{limit} is the limit value of the reduced moment.

θ is the angle between concrete compression struts and the main chord

ν_1 is a nondimensional coefficient, $\nu_1 = 0.60(1\text{-}f_{ck}/250)$;

z is the lever arm, $z = 0.9d$;

h_{fmin} is the minimum depth of flange.

3.4. Optimization based on minimum cost design

The optimum cost design of reinforced concrete T-beams under the limit state can be stated as follows:

For given material properties, loading data and constant parameters, find the design variables defined in **Table 1** that minimize the cost function defined in Eq. (4) subjected to the design constraints given in Eq. (6) through Eq. (23).

3.5. Optimization based on minimum weight design

Find the design variables that minimize total weight per unit length defined in Eq. (5), subjected to the design constraints given in Eq. (6) through Eq. (23).

3.6. Solution methodology: Generalized reduced gradient method

The objective function Eq. (4), the objective function Eq. (5) and the constraints equations, Eq. (6) through Eq.(23), together form a nonlinear optimization problem. The reasons for the nonlinearity of this optimization problem are essentially due to the expressions of the cross-sectional area, bending moment capacity and other constraints equations. Both the objective function and the constraint functions are nonlinear in terms of the design variables. In order to solve this nonlinear optimization problem, the generalized reduced gradient (GRG) algorithm is used. This algorithm was first developed in late 1960 by Jean Abadie [16] as an extension of the reduced gradient method and then since has been refined by several other researchers [17, 18]. GRG nonlinear should be selected if any of the equations involving decision variables or constraints is nonlinear.

Microsoft Excel, beginning with version 3.0 in 1991, incorporates an NLP solver that operates on values and formulas of a spreadsheet model. Version 4.0 and later include LP solver and mixed-integer programming (MIP) capability for both linear and nonlinear problems. The Microsoft Office Excel Solver tool uses several algorithms to find optimal solutions. The GRG nonlinear solving method for nonlinear optimization uses the Generalized Reduced Gradient code. The Simplex LP solving method for linear programming uses the Simplex and dual Simplex method. The Evolutionary solving method for non-smooth optimization uses a variety of genetic algorithm and local search methods. The user specifies a set of cell addresses to be independently adjusted (the decision variables), a set of formulae cells whose values are to be constrained (the constraints) and a formula cell designated as the optimization objective. The solver uses the spreadsheet interpreter to evaluate the constraint and objective functions and approximates derivatives, using finite differences. The NLP solution engine for the Excel Solver is GRG.

The generalized reduced gradient method is applied as it has the following advantages: (i) the GRG method is widely recognized as an efficient method for solving a relatively wide class of nonlinear optimization problems; (ii) the program can handle up to 200 constraints, which is suitable for reinforced ordinary and HSC beam design optimization problems; and (iii) GRG transforms inequality constraints into equality constraints by introducing slack variables. Hence all the constraints are of equality form. A more detailed description of the GRG method can be found in [19].

4. Numerical results and discussion

4.1. Design example A for reinforced HSC T-beams

The numerical example A corresponds to a high-strength concrete T-beam belonging to a bridge deck, simply supported at its ends and predesigned in accordance with provisions of EC-2 design code.

The corresponding preassigned parameters are defined as follows:

$L = 25$ m; $M_{Ed} = 1.35\ M_G + 1.5\ M_Q = 9$ MNm; $V_{Ed} = 1.35\ V_G + 1.5\ V_Q = 3.1$ MN.

w = 0.60MN/ml (the total distribution load (dead load + live load)), $\delta_{lim} = L/250 = 0.100$ m.

Input data for HSC characteristics:

C70/85; $f_{ck} = 70$ MPa; $\gamma_c = 1.5$; $f_{cd} = 46.67$ MPa; $\rho = 0.025$ MN/m^3; $E_{cm} = 40{,}743$ MPa;

$\lambda = 0.75$; $\eta = 0.90$; $\varepsilon_{cu3}(‰) = 2.7$; $\varepsilon_{c3}(‰) = 2.4$; $h_{fmin} = 0.10$ m; $f_{ctm} = 4.6$ MPa;

$\mu_{limit} = 0.329$; $\alpha_{limit} = 0.554$ for S500 and C70/85.

Input data for steel characteristics:

S500; $f_{yk} = 500$ MPa; $\gamma_s = 1.15$; $f_{yd} = f_{yk}/\gamma_s = 435$ MPa; n = 15;

S400; $f_{yk} = 400$ MPa; $\gamma_s = 1.15$; $f_{yd} = f_{yk}/\gamma_s = 348$ MPa; $f_{yd}/f_{cd} = 9.32$ for classes (S500, C70/85);

$f_{yd}/f_{cd} = 7.46$ for classes (S400, C70/85); $\mu_{limit} = 0.352$; $\alpha_{limit} = 0.6081$ for S400 and C70/85;

$E_s = 2 \times 10^5$ MPa; $p_{min} = 0.26\ f_{ctm}/f_{yk} = 0.002392$; $p_{max} = 4\%$.

Input data for units cost ratios of construction materials:

$C_s/C_c = 40$ for HSC concrete;

$C_f/C_c = 0.01$ for wood formwork;

$C_f/C_c = 0.10$ for metal formwork;

$C_f/C_c = 0.00$ in the case of the cost of the formwork is negligible.

4.1.1. Comparison between the minimum cost design and the minimum weight design of HSC T-beams

The vector of design variables including the geometric dimensions of the T-beam cross-section and the area of tension reinforcement as obtained from the standard design approach solution and the optimal cost design solution using the proposed approach is shown in **Table 2**.

Design variables vector.	Initial design	Optimal solution with minimum cost (S500, C70/85), $C_s/C_c = 40$, $C_f/C_c = 0.01$ wood formwork	Optimal solution with minimum weight
b(m)	1.20	0.86	0.52
b_w(m)	0.40	0.28	0.28
h(m)	1.40	1.58	1.56
d(m)	1.26	1.42	1.40
h_f(m)	0.15	0.11	0.10
A_S(m^2)	185x10^{-4}	161x10^{-4}	181 x10^{-4}
α	0.554	0.342	0.554
Gain		22%	47%

Table 2. Comparison of the optimal solutions with minimum weight and minimum cost design for HSC.

The optimal solutions using the minimum cost design and the minimum weight design are shown in **Table 2**.

It is shown from **Table 2** that the gain and optimum values for minimum cost design and for minimum weight design are different.

From the above results, it is clearly shown that significant cost saving of the order of 47% can be obtained using the proposed minimum weight design formulation and 22% through the use of minimum cost-design approach.

4.1.2. Parametric study

In this section, the optimal solution is obtained according to practical consideration: (i) the total depth is imposed, $h = h_{imposed;}$ (ii) the effective width of compressive flange is imposed, $b = b_{imposed;}$ (iii) the reinforcing steel is imposed, $A_s = A_{simposed;}$ and (4i) the flange depth is imposed, $h_f = h_{fimposed}$.

The gain depends on the type of formwork used. We distinguish the wood formwork: $C_f/C_c = 0.01$ and the steel formwork $C_f/C_c = 0.10$.

Further practical requirements can also be implemented, such as esthetic, architectural and limited authorized templates. The optimal solutions obtained using the particular conditions imposed are shown in **Table 3**.

From the above results, it is clearly seen that a significant cost saving between 08% and 23% can be obtained by using this parametric study.

4.1.3. Sensitivity analysis

The relative gains can be determined for various values of unit-cost ratios: $C_s/C_c = 10$; 20; 30; 40; 50; 60; 70; 80; 90; 100 for a given unit cost ratio $C_f/C_c = 0.01$

The corresponding results are reported in **Table 4** and represented in **Figure 2**.

Optimal solution with.	Gain (%)
Classes(S500, C70/85); $C_s/C_c = 40$; $C_f/C_c = 0.01$wood formwork	22
Classes(S500, C70/85); $C_s/C_c = 40$; $C_f/C_c = 0.10$ steel formwork	19
Classes(S500,C70/85) and $C_f/C_c = 0$ the cost of the formwork is negligible	23
Classes(S400,C70/85); $C_s/C_c = 40$; $C_f/C_c = 0.01$wood formwork	08
Imposed height h = 1.70 m; S500 and C70/85	21
Imposed width b = 1.00 m; S500 and C70/85	22
Imposed reinforcement $A_s \leq 0.0150$ m^2; S500 and C70/85	22
Imposed flange depth $h_f = 0.10$ m; S500 andC70/85	22

Table 3. The variation of relative gain with particular conditions imposed such as the HSC T-beam dimensions and reinforcing steel.

(S500; C70/85) $C_f/C_c = 0.01$	Gain (%)
10	33
20	27
30	24
40	22
50	22
60	22
70	23
80	24
90	26
100	27

Table 4. Variation of relative gain in percentage (%) versus unit cost ratio C_s/C_c for a given cost ratio $C_f/C_c = 0.01$.

It is shown in **Table 4** and **Figure 2** that the relative gain decreases for increasing values of the unit cost ratio C_s/C_c, stabilizes around an average value for $40 \leq C_s/C_c \leq 60$ and then increases significantly beyond this average value for a given cost ratio $C_f/C_c = 0.01$.

The relative gains can be determined for various values of unit cost ratios: $C_f/C_c = 0.01$; 0.02; 0.03; 0.04; 0.05; 0.06; 0.07; 0.08; 0.08; 0.09; 0.10 for a given unit cost ratio $C_s/C_c = 40$.

The corresponding results are reported in **Table 5** and presented in **Figure 3**.

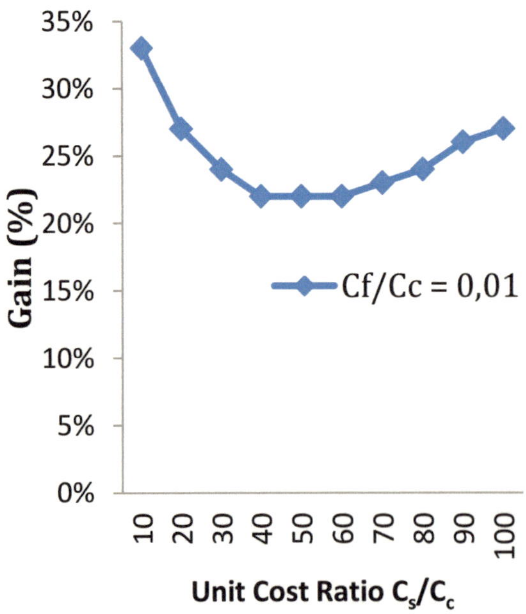

Figure 2. Variation of relative gain in percentage (%) versus unit cost ratio C_s/C_c for a given cost ratio $C_f/C_c = 0.01$.

(S500; C70/85) C_s/C_c = 40 C_f/C_c	Gain (%)
0.01	22
0.02	21
0.03	21
0.04	20
0.05	20
0.06	19
0.07	19
0.08	19
0.09	19
0.10	18

Table 5. Variation of relative gain in percentage (%) versus unit cost ratio C_f/C_c for a given cost ratio C_s/C_c = 40.

From **Table 5** and **Figure 3**, the gain decreases monotonically with the increase of unit cost ratio C_f/C_c for a given cost ratio C_s/C_c = 40.

4.2. Design example B for reinforced ordinary concrete T-beams

The numerical example B corresponds to a concrete T-beam belonging to a pedestrian deck, simply supported at its ends and predesigned in accordance with the provisions of EC-2 design code.

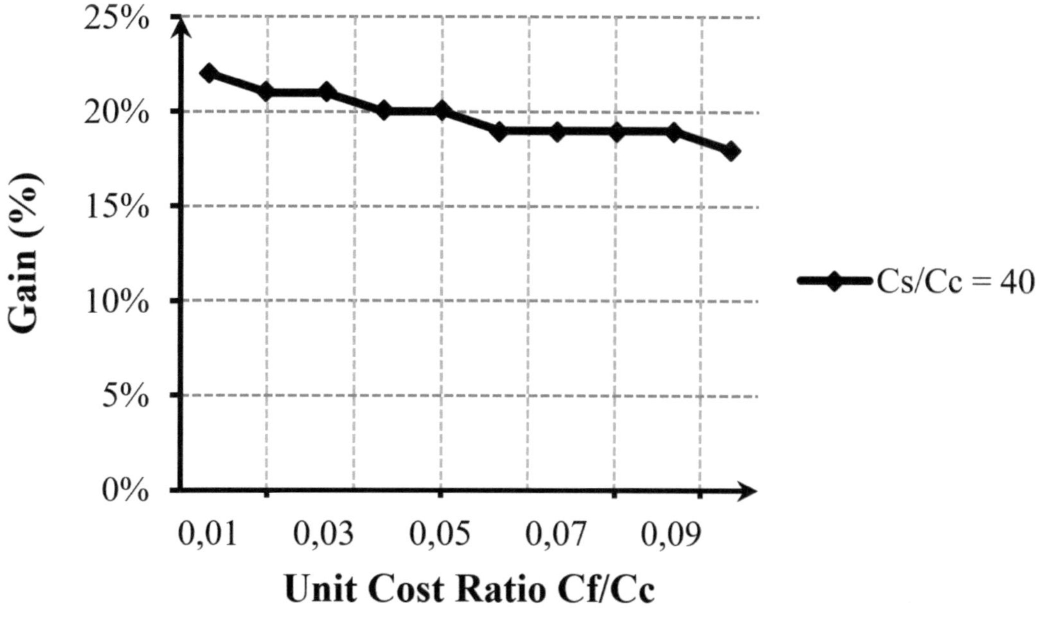

Figure 3. Variation of relative gain in percentage (%) versus unit cost ratio C_f/C_c for a given cost ratio C_s/C_c = 40.

The preassigned parameters are defined as follows:

L = 20 m; M_{Ed} = 5MNm; V_{Ed} = 1.1MN; w = 0.043MN/ml; δ_{lim} = L/250 = 0.080 m.

Input data for ordinary concrete characteristics:

C20/25; f_{ck} = 20 MPa; γ_c = 1.5;f_{cd} = 11.33 MPa; ρ = 0.025MN/m^3; E_{cm} = 30,000 MPa;

λ = 0.80; η =1.00; ε_{cu3}(‰) = 2; ε_{c3}(‰) = 3.5; h_{fmin} = 0.15 m; f_{ctm} = 2.20 MPa; n = 15;

μ_{limit} = 0.372; α_{limit} = 0.6167 for S500 and C20/25.

μ_{limit} = 0.392; α_{limit} = 0.6680 for S400; and C20/25.

Input data for steel characteristics:

S400; f_{yk} = 400 MPa; γ_s = 1.15; $f_{yd} = f_{yk}/\gamma_s$ = 348 MPa;

$E_s = 2 \times 10^5$ MPa; $p_{min} = 0.26\ f_{ctm}/f_{yk}$ = 0.00143; p_{max} = 4%;

f_{yd}/f_{cd} = 30.71 for classes (S400, C20/25);

f_{yd}/f_{cd} = 38.39 for classes (S500, C20/25).

Input data for units cost ratios of construction materials:

C_s/C_c = 30 for ordinary concrete.

C_f/C_c = 0.10 for metal formwork.

C_f/C_c = 0.01 for wood formwork.

4.2.1. Comparison between the minimum cost design and the minimum weight design of ordinary concrete T-beams

The optimal solutions using the minimum weight design and the minimum cost design are shown in **Table 6**.

It is shown in **Table 6** that the gain and the optimum values for minimum weight design and for minimum cost design are different.

From the above results, it is clearly shown that a significant cost saving of the order of 23% can be obtained using the proposed minimum weight design formulation and 14% through the use of the minimum cost design approach.

4.2.2. Parametric study

In this section, the optimal solution is obtained through the considerations: (i) one of the dimensions of HSC T-section is imposed, h = 1.50 m; (ii) the imposed reinforcing steel A_s = 120×10^{-4} m^2; (iii) imposed web width b_W = 0.30 m; and (iv) imposed relative depth of compressive concrete zone α = 0.6000

Design variables vector	Initial design, C20/25 & S400	Optimal solution with minimum weight, C20/25 & S400	Optimal solution with minimum cost, C20/25 & S400
b(m)	1.20	1.30	1.25
b_w(m)	0.40	0.28	0.29
h(m)	1.60	1.57	1.60
d(m)	1.44	1.41	1.44
h_f(m)	0.14	0.17	0.16
A_S(m^2)	125×10^{-4}	123×10^{-4}	122×10^{-4}
α	0.668	0.668	0.668
C	1.171		1.0281
Gain		23%	14%

Table 6. Comparison of the optimal solutions with minimum weight and minimum cost design.

Optimal solution with	Gain (%)
$f_{yd}/f_{cd} = 30.71$; $C_s/C_c = 30$; $C_f/C_c = 0.01$ wood formwork, C20/25 & S400	14
$f_{yd}/f_{cd} = 38.39$; $C_s/C_c = 30$; $C_f/C_c = 0.01$wood formwork, C20/25 & S500	09
$f_{yd}/f_{cd} = 30.71$; $C_s/C_c = 30$; $C_f/C_c = 0.00$; C20/25 & S400	15
Imposed web with $b_w = 0.30$ m; $f_{yd}/f_{cd} = 30.71$; $C_s/C_c = 30$; $C_f/C_c = 0.01$; C20/25 & S400	13
Imposed reinforcement$A_s \leq 0.0120$ m^2; $f_{yd}/f_{cd} = 30.71$; $C_s/C_c = 30$; $C_f/C_c = 0.01$; C20/25 & S400	14
Imposed height h = 1.50 m; $f_{yd}/f_{cd} = 30.71$; $C_s/C_c = 30$; $C_f/C_c = 0.01$; C20/25 & S400	11
Imposed relative depth$\alpha = 0.600$; $f_{yd}/f_{cd} = 30.71$; $C_s/C_c = 30$; $C_f/C_c = 0.01$; C20/25 & S400	14

Table 7. Variation of relative gain with particular conditions imposed such as the T-beam dimensions, reinforcing steel and weight.

(S400; C20/25) $C_f/C_c = 0.01$	Gain (%)
10	18
20	16
30	14
40	13
50	12
60	12
70	12
80	11
90	11
100	11

Table 8. Variation of relative gain in percentage (%) versus unit cost ratio C_s/C_c for a given cost ratio $C_f/C_c = 0.01$.

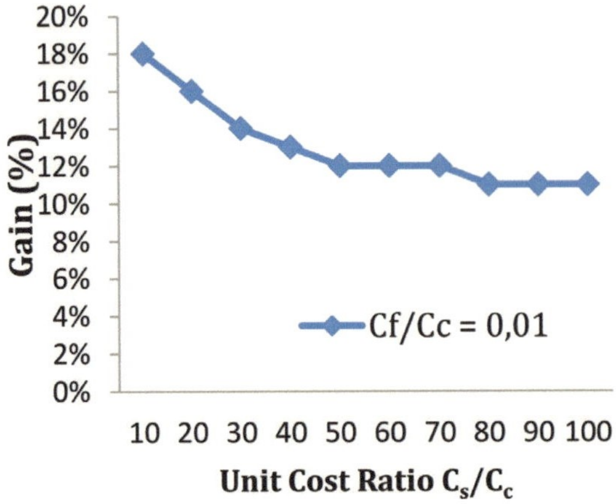

Figure 4. Variation of relative gain in percentage (%) versus unit cost ratio C_s/C_c for a given cost ratio $C_f/C_c = 0.01$.

(S400; C20/25) $C_s/C_c = 30$ C_f/C_c	Gain (%)
0.01	14
0.02	14
0.03	13
0.04	13
0.05	13
0.06	12
0.07	12
0.08	12
0.09	12
0.1	12

Table 9. Variation of relative gain in percentage (%) versus unit cost ratio C_f/C_c for $C_s/C_c = 30$.

Further practical requirements can also be implemented, such as esthetic, architectural and limited authorized template.

The optimal solutions obtained using the particular conditions imposed are shown in **Table 7**.

From the above results, it is clearly seen that a significant cost saving between 09 and 15% can be obtained by using this parametric study.

4.2.3. Sensitivity analysis

The relative gains can be determined for various values of the unit cost ratios: $C_s/C_c = 10$; 20; 30; 40; 50; 60; 70; 80; 90; 100 for a given unit cost ratio $C_f/C_c = 0.01$

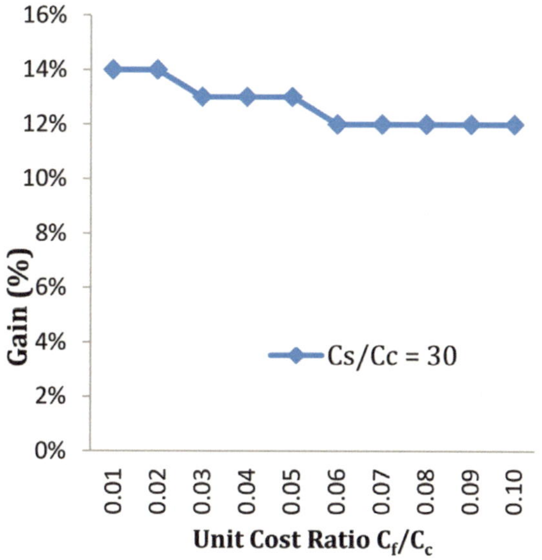

Figure 5. Variation of relative gain in percentage (%) versus unit cost ratio C_f/C_c for a given cost ratio $C_s/C_c = 30$.

The corresponding results are reported in **Table 8** and presented graphically in **Figure 4**.

It is shown in **Table 8** and **Figure 4** that the relative gain decreases for increasing values of the unit cost ratio C_s/C_c for a given value of $C_f/C_c = 0.01$.

The relative gains can be determined for various values of the unit cost ratios: $C_f/C_c = 0.01$; 0.02; 0.03; 0.04; 0.05; 0.06; 0.07; 0.08; 0.08; 0.09; 0.10 for a given unit cost ratio $C_s/C_c = 30$.

The corresponding results are reported in **Table 9** and illustrated graphically in **Figure 5**.

From **Table 9** and **Figure 5**, the gain decreases monotonically with the increase of unit cost ratio C_f/C_c for a given value of $C_s/C_c = 30$.

5. Conclusions

The following important conclusions are drawn on the basis of this chapter:

- The problem formulation of the optimal cost design of reinforced concrete T-beams can be cast into a nonlinear programming problem; the numerical solution is efficiently determined using the GRG method in a space of only a few variables representing the concrete cross-section dimensions.
- The space of feasible design solutions and the optimal solutions can be obtained from a reduced number of independent design variables.
- The optimal values of the design variables are only affected by the relative cost values of the objective function and not by the absolute cost values.

- The optimal solutions are found to be insensitive to changes in the shear constraint. Shear constraint is not usually critical in the optimal design of reinforced concrete T-beams under bending and thus can be excluded from problem formulation.
- The observations of optimal solution results reveal that the use of optimization based on the optimum cost design concept may lead to substantial savings in the amount of construction materials to be used in comparison to classical design solutions of reinforced concrete T-beams.
- The objective function and the constraints considered in this chapter are illustrative in nature. This approach based on nonlinear mathematical programming can be easily extended to other sections commonly used in structural design. More sophisticated objectives and considerations can be readily accommodated by suitable modifications of the optimal cost design model.
- In this chapter, we have included the additional cost of formwork which makes a significant contribution to the total costs. This integration is important for an economical approach to design and manufacture.
- The suggested methodology for optimum cost design is effective and more economical compared to the classical methods. The results of the analysis show that the optimization process presented herein is effective and its application appears feasible.
- The comparison of optimal solutions for minimum cost and minimum weight shows that the construction cost affects significantly the optimal sizes. Not only do we use the mass but the cost as objective function as well which contains the material and construction provision costs. The difference is caused by construction details costs.

Appendix

List of symbols

The following symbols are used in this chapter:

C20/25	Class of ordinary concrete
C70/85	Class of HSC
S400	Grade of steel
S500	Grade of steel
f_{ck}	Characteristic compressive cylinder strength of ordinary or HSC at 28 days
f_{ctm}	Tensile strength of concrete
f_{cd}	Design value of concrete compressive strength
γ_c	Partial safety factor for concrete

η	Design strength factor
λ	Compressive zone depth factor
ε_{c3}	Strain at the maximum stress for the rectangular stress distribution compressive concrete
ε_{cu3}	Ultimate strain for the rectangular stress distribution compressive concrete design stress–strain relation
f_{yk}	Characteristic elastic limit for steel reinforcement
γ_s	Partial safety factor for steel
f_{yd}	Design yield strength of steel reinforcement
ε_{yd}	Elastic limit strain
E_s	Young's elastic modulus of steel
E_{cm}	Modulus of elasticity of concrete
p_{min}	Minimum steel percentage
p_{max}	Maximum steel percentage
α_{limit}	Limit value of relative depth of compressive concrete zone
μ_{limit}	Limit value of reduced moment
L	Beam span
w	The total distribution load (dead load+ live load)
V_G	Maximum design shears under dead loads
V_Q	Maximum design shears under live loads
$V_{Rd,max}$	Maximum resistant shear force
V_{Ed}	Ultimate shear force
$M_{Rd, max}$	Maximum resisting moment
M_{Ed}	Ultimate bending moment
M_G	Maximum design moments under dead loads
M_Q	Maximum design moments under live loads
F_s	Resultant tensile internal force for steel
F_c	Resultant compressive internal force for HSC
n	Ratio of the modulus of elasticity of steel to that of concrete
b	Effective width of compressive flange

b_w	Web width
h	Total depth
h_f	Flange depth
d	Effective depth
d_s	Effective cover of reinforcement
As	Area of reinforcing steel
h_{fmin}	Minimum depth of flange
δ_w	The mid-span deflection of simply supported beam under distribution load w
δ_{lim}	Limit deflection
θ	Angle between concrete compression struts and the main chord
ν_1	A nondimensional coefficient; $\nu_1 = 0.60(1\text{-}f_{ck}/250)$
z	Lever arm, $z = 0.9d$
ρ	Density of the reinforced concrete T-beams
W	Unit weight per unit length of the reinforced concrete T- beams
C_0/L	Total cost per unit length of T-beam
C_s	Unit cost of reinforcing steel
C_c	Unit cost of concrete
C_f	Unit cost of formwork

Author details

Fedghouche Ferhat

Address all correspondence to: ferfed2002@yahoo.fr

École Nationale Supérieure des Travaux Publics (ENSTP), Département Infrastructures de Base (DIB), Laboratoire des Travaux Publics ingénierie de Transport et Environment (LTPiTE), Algiers, Algeria

References

[1] Babu Narayan KS, Venkataramana K. Shape optimization of steel reinforced concrete beams. Computers and Concrete. 2007;**4**:317-330. DOI: 10.12989/cac.2007.4.4.317

[2] Ozturk HT, Durmus A. Optimum design of a reinforced concrete beam using artificial bee colony algorithm. Computers and Concrete. 2012;**10**:295-306. DOI: 10.12989/cac.2012.10.3.295

[3] Bordignon R, Kripka M. Optimum design of reinforced concrete columns subjected to uniaxial flexural compression. Computers and Concrete. 2012;**9**:345-358. DOI: 12989/cac.2012.9.5.327

[4] Medeiros GF, Kripka M. Structural optimization and proposition of pre-sizing parameters for beams in reinforced concrete buildings. Computers and Concrete. 2013;**11**:253-270. DOI: 10.12989/cac.2013.11.3.253

[5] Rashid MA, Mansur MA. Reinforced high strength concrete beams in flexure. ACI Structural Journal. 2005;**03**:462-471

[6] Ozbay E, Oztas A, Baykasoglu A. Cost optimization of high strength concretes by soft computing techniques. Computers and Concrete. 2010;**7**:221-237. DOI: 10.12989/cac.2010.7.3.221

[7] Nocedal J, Wright SJ. Numerical optimization. Second ed. New York: Springer Verlag; 2006

[8] Jiang P, Zhu M, XLJ. An optimization algorithm for minimizing weight of the composite beam. Advances in Intellegent and soft.Computing. 2012;**135**:769-775. DOI: 10.1007/978-3-642-27708-5_106

[9] Kaveh A, Kalateh-Ahani M, Fahimi-Farzam M. Constructability optimal design of reinforced concrete retaining walls using a multi-objective optimization of foundation using global-local gravitational search algorithm. Structural Engineering and Mechanics. 2013;**50**:257-273. DOI: 10.12989/sem.2014.50.3.257

[10] Fedghouche F, Tiliouine B. Minimum cost design of reinforced concrete T-beams at ultimate loads using Eurocode 2. Engineering Structures. 2012;**42**:43-50. DOI: 10.1016/j.engstruct.2012.04.008

[11] Abbasnia R, Shayanfar M, Khodam A. Reliability-based design optimization of structural systems using a hybrid genetic algorithm. Structural Engineering and Mechanics. 2014;**52**:1099-1120. DOI: 10.12989/sem.2014.52.6.1099

[12] Fedorik F, Kala J, Haapala A, Malaska M. Use of design optimization techniques in solving typical structural engineering related design optimization problems. Structural Engineering and Mechanics. 2015;**55**:1121-1137. DOI: 10.12989/sem.2015.55.6.1121

[13] Eurocode 2. Design of concrete structures-part 1–1: general rules for buildings.EN 1992–1-1:2004, European Committee for Standardization (CEN), Brussels; 2004

[14] Pratt DJ. Fundamentals of Construction Estimating. 3rd ed. Australia: Delmar cengage Learning; 2011

[15] Davis L. Spon's Architects' and Builders' Price Book 2011. 136th ed. Taylor & Francis Ltd.; 2010

[16] Abadie J, Carpentier J. In: Fletcher R, editor. Generalization of the Wolfe Reduced Gradient Method to the Case of Nonlinear Constraints. Optimization (Sympos., Univ. Keele, Keele, 1968). London: Academic Press; 1969

[17] Lasdon S, Warren AD, Jain A, Ratner M. Design and testing of a generalized reduced gradient code for nonlinear programming. ACM Trans. Math. Software. 1978;**4**(1):34-50

[18] Lasdon LS, Waren AD. Generalized reduced gradient method for linearly and nonlinearly constrained programming. In: Greenberg HJ, editor. Design and Implementation of Optimization Software. the Netherlands: Sijthoff and Noordhoff; 1978. pp. 363-397

[19] Yeniay O. A comparative study on optimization methods for the constrained nonlinear programming problems. Mathematical Problems in Engineering. 2005;(2):165-173. DOI: 10.1155/MPE.2005.165

Chapter 8

Improved Dielectric Properties of Epoxy Nano Composites

Rashmi Aradhya and Nijagal M. Renukappa

Additional information is available at the end of the chapter

http://dx.doi.org/10.5772/intechopen.78057

Abstract

Epoxy-based nanodielectrics with 2, 5 and 7 wt.% of organically modified montmorillonite clay (oMMT) were prepared using high shear melt mixing technique. The interface of oMMT and epoxy of the nanodielectrics play a very important role in improving electrical, mechanical, thermal and wear properties. Therefore detailed study on the interfacial effects of filler-matrix has been investigated for understanding the chemical bonding using Fourier transform infrared spectroscopy (FTIR) and the cross linking between polymer and filler was studied using glass transition temperature (T_g) through differential scanning calorimetry (DSC). Further, positron annihilation lifetime spectroscopy (PALS) was used to determine precise and accurate value of free volume of the nanodielectrics. The interaction between the nanoparticles and polymer chains has a direct bearing on dielectric strength characteristics of the epoxy-oMMT nanocomposite system and accordingly, the ac dielectric strength of the nanodielectrics increases with the addition of oMMT into epoxy up to 5 wt.% and further increase in filler loading (7 wt.%) causes decrease in ac dielectric strength.

Keywords: Fourier transform infrared spectroscopy, positron annihilation life time spectroscopy, dielectric strength, glass transition temperature, differential scanning calorimetry, epoxy-oMMT nanocomposite

1. Introduction

The interface of surface functionalized nano-clay (oMMT) filler and polymer matrix (epoxy) of the polymer nanocomposites play a very important role in improving the electrical, thermal and mechanical properties. Therefore, detailed studies on the interfacial effects of filler-matrix on several properties have been investigated. The chemical bonding established between epoxy and oMMT nanofiller has been investigated using Fourier transform infrared spectroscopy (FTIR). The cross linking between polymer and nanofiller was measured to determine the glassy state of the nanocomposite called glass transition temperature (T_g) by using differential scanning calorimetry (DSC). Further, the positron annihilation spectroscopy (PALS) has been utilized to determine free volume as outlined in the multi-core model [1]. Many researchers have theoretically estimated the free volume of nanocomposites and there is no experimental data on the evaluation of free volume. In the present work, PALS has been used in the accurate evaluation of free volume. A brief explanation of nanocomposite interface dynamics, free volume estimation and the effect of intermolecular interactions and hydrogen bonding are discussed. The effect of these results on electrical property such as dielectric strength (DES) at room temperature was studied.

1.1. Multi-core interface model

The interaction of nanoparticles with the surrounding polymer matrix by means of three layers is described by Multi-core model [1] as shown in **Figure 1**. It consists of (i) Primary layer also referred as bonded layer, (ii) Secondary layer as referred as called bound layer, (iii) tertiary layer also referred as called loose layer, and the next fourth layer which overlaps all the above three layers called electric double layer. Primary layer represents a type of transition layer which is firmly attached to the carbonless nanofiller or inorganic nanofiller and carbon based organic base matrix polymer by compatabilizer or hardner. Secondary or bound layer addressed as interfacial layer or region consists of a region or area of layer of polymer chains

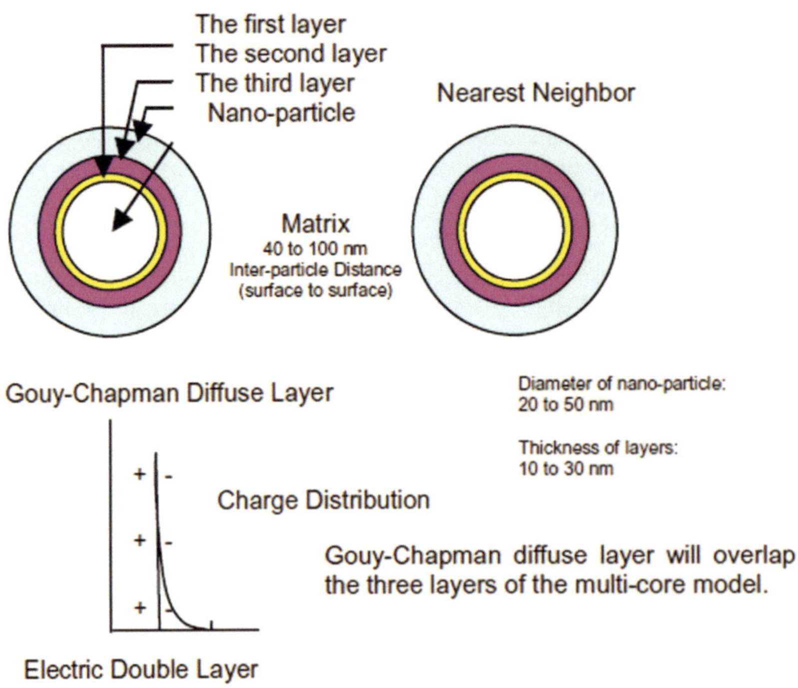

Figure 1. Multi-core model for nano-particle-polymer interfaces (source: Toshikatsu Tanaka and co-authors [1]).

strongly bound to the primary layer and outer surface of nanofiller. Thickness of the layer varies between 2 and 9 nm. This value mainly relies on the strength of interaction between organic polymer and nanofiller or nanoparticles. Stronger the interaction, the larger will be the bound polymer fraction. This may correspond to a stoichiometrically cross-linked layer.

The loose layer is a region which is loosely coupled, at the same time interacts with the bound region. It is generally considered that the loose layer has different chain conformation, chain mobility, and even free volume in the polymer matrix. It may also consist of a less stoichiometrically cross-linked layer. In addition, interfacial structures obtained from chemistry, Columbic interaction is superimposed, when dielectric and electrical insulation properties are investigated [1]. The nanoparticle may be charged either positively or negatively. When a polymer has mobile charge carriers, they are distributed in the interface in such a manner that the charge carriers with the opposite polarity are diffused outward from the contact surface to the Debye shielding length that corresponds to the Gouy-Chapman diffuse layer in which charge decays exponentially with distance, in accordance with Born approximation [1]. Debye shielding length is calculated approximately as 30 nm [1].

The thickness of these layers may varies from 1 nm to, several tens nm for the bonded, bound and loose layers respectively. It is not clearly known whether the thickness of the loose layer is the same as that of the Gouy-Chapman diffuse layer. It may appear that the latter might extend over the former. Therefore, far-field effect must be involved in mesoscopic interactions in the loose layer or the diffuse Gouy-Chapman layer, and it is expected to cause some combined effect among neighboring nanoparticles. Macroscopic phenomena and parameters are different from polymer to polymer and polymer with filler particles due to the relative differences in its thickness and interaction strengths in the multi-core model with the far-field effect.

2. Experimental studies

2.1. Matrix and fillers

The Bisphenol A diglycidyl ether based Epon 828 epoxy resin (DGEBA) with epoxy equivalent weight (EEW) of 188 g/mol and curing agent such as Epikure W which is chemically called diethyl toluene diamine (DETDA) with an amine hydrogen equivalent weight of 45 g/mol were used in the present work. These materials were supplied by M/s. Miller-Stephenson Chemical Company, USA. The nanoclay used in the present work is called as Nanomer 1.30E supplied by M/s. Nanocor, USA. This nanoclay was surface treated with surface functionalizer namely octadecylamine mainly used for uniform dispersion of nanoparticle epoxy resin polymer. This surface functionalized nanoclay is called organically modified montmorillonite clay represented as oMMT.

2.2. Fabrication process of nanocomposite

One of major challenges in the processing of nanocomposites is the non-uniform mixing of curing agent. Non uniform mixing of nanofiller, resin, and the curing agent or hardner may also results in an improper curing. The schematic diagrams of processing method and curing cycle are shown in **Figures 2** and **3** respectively.

Steps for processing of epoxy-oMMT nanocomposites:

Step1: The viscosity of Epon 828 resin was high at room temperature; and therefore, it was difficult to mix. In order to reduce the viscosity before mixing, the resin was preheated in an oven at 60°C for about 2 hours. The filler was dried in an oven at 100°C for 24 hours.

Step2: After reducing the viscosity, a known weight of epoxy resin was taken.

Step3: The required weight of the oMMT and curing agent were added to the epoxy resin.

Step4: Epoxy resin, oMMT, and curing agent mixture were mixed using IKA high shear mixer (T-T18 ULTRA TURRAX Basic) at a speed of 24,000 RPM for 45 minutes.

Step5: After mixing, degassing was carried out in a vacuum oven for 45 minutes.

Step6: The, the mixture was then transferred into aluminum molds and degassed again for 30 minutes. After degassing, the aluminum molds were placed in an oven, and the materials were cured based on the time–temperature curing cycle shown in **Figure 3**. The dimensions of cured sheet of epoxy-oMMT nanocomposites used for the investigation had an area of 200 mm × 200 mm and thickness of 3 mm.

2.3. Measurements

2.3.1. Interface dynamics

(i) Fourier transform infrared spectroscopy (FTIR)

Nanoparticles chemistry and chemical bonding type that existed between polymer and nanoparticles of cured polymer nanocomposites in the present study were characterized through FTIR measurements.

(ii) Differential scanning calorimeter (DSC)

The nanocomposites melting and glass transition temperature were calculated using DSC model 821 of Thermal Analysis instruments.

(iii) Positron annihilation lifetime spectrometer (PALS)

The PAL spectra of the nanocomposites have been traced by means of positron lifetime spectrometer with time resolution of 220 picoseconds.

(iv) Dielectric strength (DES)

The electrical breakdown measurements were carried out using HV AC Test Set of M/s. W.S. Test Systems Pvt. Ltd., Bangalore.

2.4. Experimental techniques

The bonding of nanoparticles with the base polymer and the chemical nature of cured polymer material were characterized through FTIR measurements using Perkin Elmer make, model spectrum-GX FT-IR as per ASTM D 7214-07a.

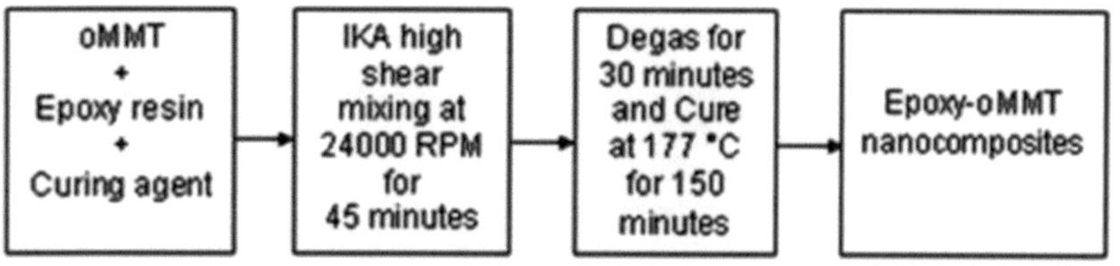

Figure 2. Processing of epoxy-oMMT nanocomposites.

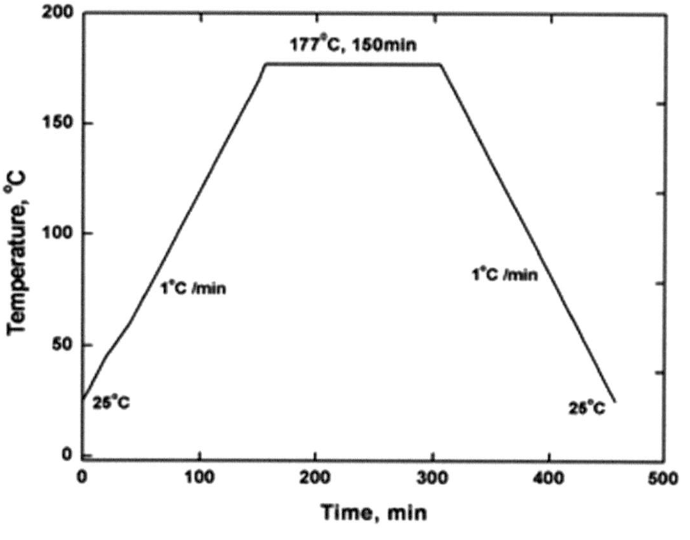

Figure 3. Curing cycle for epoxy-oMMT nanocomposites.

The DSC Model 821 of Thermal Analysis instrument was operated in nitrogen atmosphere to determine the glass transition temperature as per ASTM D-3428-99. The PALS in pure epoxy resin and oMMT filled nanocomposites were recorded using the positron lifetime spectrometer with a time resolution of 220 ps [2]. The PAL spectra obtained were analyzed by employing the computer software PATFIT [3]. This software decomposes a PAL spectrum into three discs. The lifetime component τ_2 with intensities I_2 is due to trapping of positrons at the defects. The longest life components τ_3 with intensity I_3 is due to pick-off annihilation of the o-Ps from the free volume sites present mainly in the amorphous regions of the polymer matrix. Nakanishi and co-authors [4] proposed equation 1, which is referred to the previous research works of Tao [5] and Eldrup [6] utilized to compute radius(R) of the free volume cell from the noted values of τ_3 (o-Ps lifetime).

$$\frac{1}{\lambda_3} = \tau_3 = 0.5\left[1-\frac{R}{R+\Delta R}+\frac{1}{2\pi}\sin\left(\frac{2\pi R}{R+\Delta R}\right)\right]^{-1} \quad (1)$$

Here, ΔR is the fitting parameter and has been found to be 1.656 Angstrom (Å) for solid molecular media [7]. The free volume size is evaluated as $V_f = (4/3)\,\pi R^3$.

The fractional free volume or the free volume content (F_v) is calculated as $Fv = CV_f I_3$, where, $C = 0.00018$ nm^3 [7, 8].

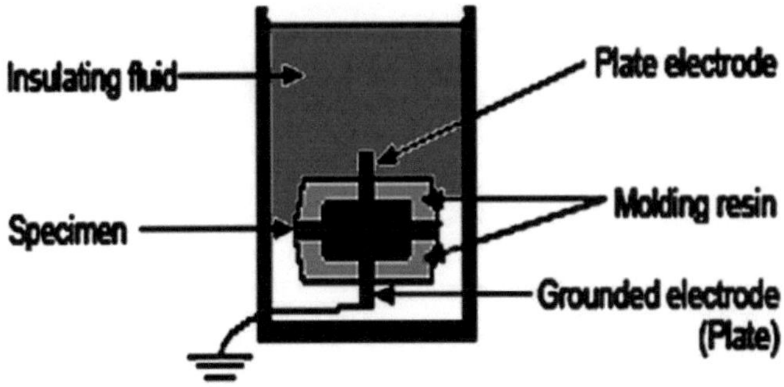

Figure 4. Insulation breakdown test method by needle-plate electrode geometry.

The electrical breakdown voltage was measured at 300.15 K as per ASTM D-149 as shown in **Figure 4**.

Electrical breakdown voltage of nanocomposite insulator was documented. The dielectric breakdown strength (DES) was computed using,

$$E = \frac{V}{t} \text{KV/mm} \quad (2)$$

here, V = Electrical breakdown voltage in kilo volts,

t = thickness of nanocomposite sample in m.

3. Results and discussion

3.1. FTIR spectrum of pure epoxy

FTIR spectrum of pure epoxy is shown in **Figure 5**. In the below figure, peaks corresponding to the presence of functional groups in the epoxy system and are listed in **Table 1**.

The epoxy resin contains several polar groups which can interact with the surface –OH groups on the nanoparticles through hydrogen bonding rather easily, similar to the bonding between the –OH groups and H_2O molecules in the atmosphere. An uncured DGEBA epoxy resin is polar in nature and contains two epoxide groups at both ends.

3.2. Hydrogen bond formation in polymer and nanoparticle

In the fabrication process of the epoxy nanocomposites, the filler (oMMT) is first mixed with DGEBA based epoxy subsequently the addition of hardner (DETDA) into the epoxy and oMMT nanoparticle mix to initiate the curing process. When DETDA hardner is added to the epoxy and oMMT particle mix, the epoxide group open up and form hydrogen bond with the free –OH groups on the nanoparticle surface in addition to reacting with amine groups of the hardner as shown in **Figure 6**.

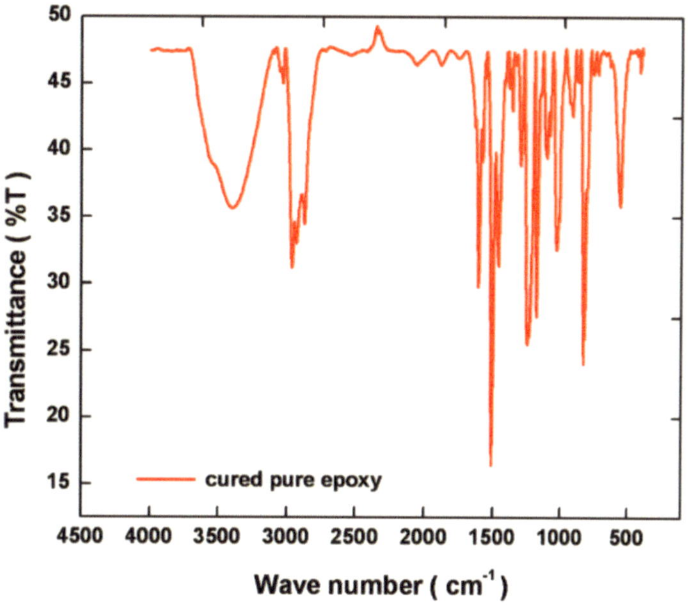

Figure 5. FTIR spectra of pure epoxy.

Wave number (cm^{-1})	Functional groups
~3399	OH groups
~3036	Corresponds to the C–H stretch in aromatics
~2965	Corresponds to asymmetrical C–H stretch of –CH_3 group
~2932	Corresponds to asymmetrical C–H stretch of –CH_2 group
~1608	Corresponds to C–C stretching vibration in aromatic
~1582	
~1508	
~1458	
~1298	Corresponds to asymmetrical –CH2 deformation
~1247	Corresponds to asymmetrical aromatic C–O stretch
~1181	Corresponds to asymmetrical aliphatic C–O stretch
~1085	Corresponds to symmetrical aromatic C–O stretch
~916	Corresponds to epoxide ring vibrations
~874	Corresponds to –CH out of plane deformation in aromatic
~560	

Table 1. FTIR peaks corresponding to functional groups in pure epoxy.

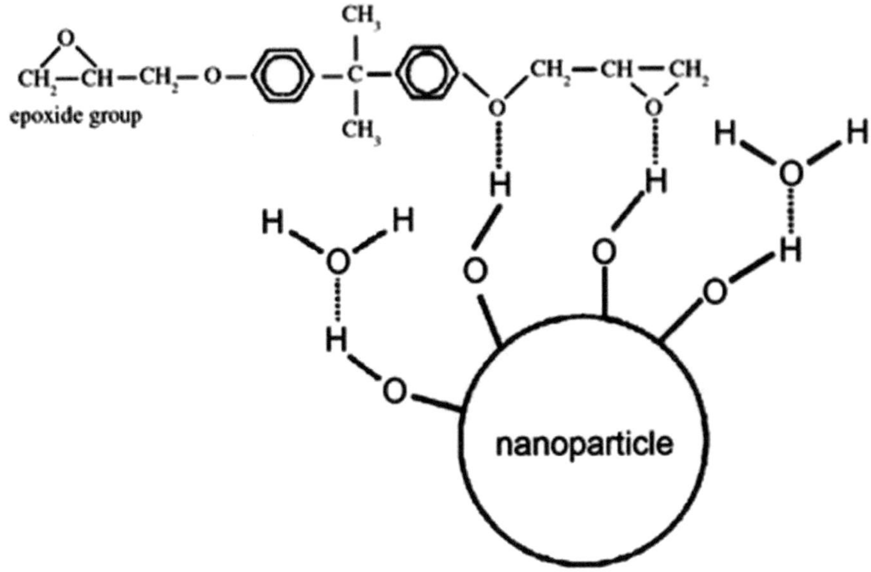

Figure 6. Hydrogen bonding between –OH groups on the nanoparticle surface and epoxide groups in epoxy during curing.

Since, there is abundantly free –OH groups are available on the nanoparticle surface, the hydrogen bonded epoxy segments will be more in number nearer to the nanoparticle surface. Development of the hydrogen bonds at interface region of epoxy and nanofiller is due to strongly bonded first nanolayer and tightly bounded second nanolayer of nanofiller and epoxy segments. This is in accordance with multi core model [1].

3.3. FTIR spectra of epoxy-oMMT nanocomposites

The FTIR spectrum of epoxy-oMMT nanocomposites is shown in **Figure 7**. The FTIR spectra clearly show that the incorporation of the filler particles leads to formation of hydrogen bonds. The interaction mechanism between nanofiller particles and base epoxy is the formation of hydrogen bond with free surface hydroxyl (–OH) groups present on the nanofiller particle outer surface and the base polymer epoxy. One of the major understand with the hydrogen bonding is that they will not exhibit any such new peaks in FTIR spectra.

The band at 3444 cm^{-1} correspond to the stretching vibration of hydroxyl (–OH) groups attached to the nanoparticle surface. The –OH groups may be attached on the surface of nanofiller particles as free –OH groups, or as –OH groups attached to absorbed H_2O molecules. When the band appears at 1638 cm^{-1}, this corresponds to the bending vibrations of absorbed H_2O molecules. The presence of the FTIR peaks at the above mentioned wavelengths reports the presence of hydroxyl (-OH) groups on the nanofiller surfaces. This result obtained through FTIR measurements in the present work have been well supported by several researchers [9–11].

The reason for the presence of H_2O molecules on the surfaces of oMMT particles are affinity of H_2O molecules present in atmosphere to get bonded to the surface OH groups on the oMMT particles through hydrogen bonding. H_2O is a polar particle or molecule and O_2 attached to the

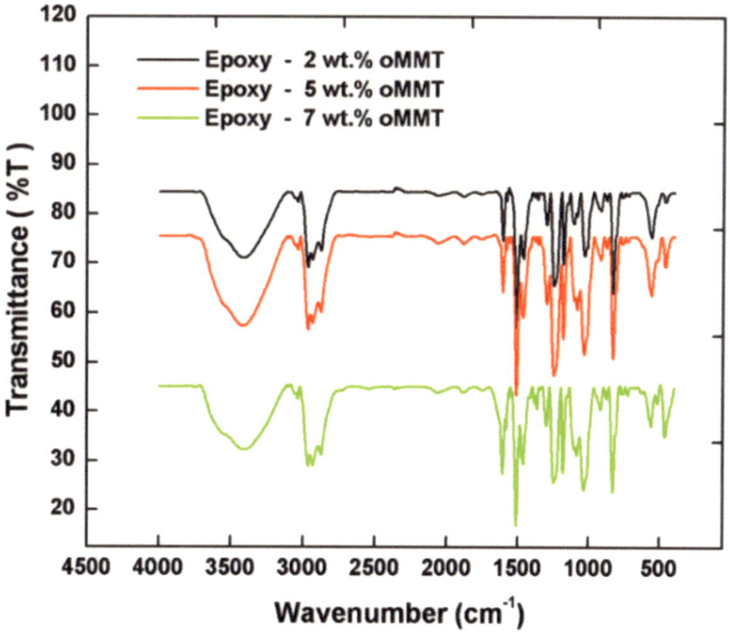

Figure 7. FTIR spectra of epoxy-oMMT nanocomposites.

H_2O molecule is highly prone to hydrogen bonding with free hydroxyl groups present on the OMMT particle surfaces as shown in **Figure 7**. Hence, there will be a tendency for all the –OH groups present in the oMMT nanoparticle surface to form hydrogen bonds with H_2O molecules in the atmosphere which may result in the formation of a H_2O layer on the surface of the OMMT particles. When such oMMT particles having high concentration of surface H_2O molecules are introduced into epoxy, the H_2O molecules will tend to influence the thermal and electrical properties of the epoxy based nanocomposites [12]. Although, it is very difficult to fully remove the formation of hydrogen bonds between the surface –OH groups and H_2O molecules, the experiments carried out have to be controlled to reduce the formation of a H_2O layer on the surface of the nanoparticles. One of the important processes which has been adopted in the present work is to adequately dry the OMMT particles before they are adding into the polymer matrix.

With reference to the **Figure 7**, the width of the peak, intensity of the peak and –OH infrared peak positions are influenced by the existence of hydrogen bonding in the oMMT nanofiller and epoxy matrix. However, it is well known that for –OH bands i)width of peak, and ii) intensity of the peak increases as the incorporation of oMMT particles in epoxy resin increases which is considered as sign of adding up of hydrogen bonds in the nanofilled filled composites.

3.4. Analysis of the glass transition temperature (T_g)

The variations in T_g of the epoxy nanocomposites are shown in the **Figure 8**. It is noted from the figure, about the T_g value of pure epoxy resin as 442.15 K. It is also noted from the plot that inclusion of 2 wt.% oMMT, into epoxy results in increased value of T_g by 278.15 K as compared to pure epoxy.

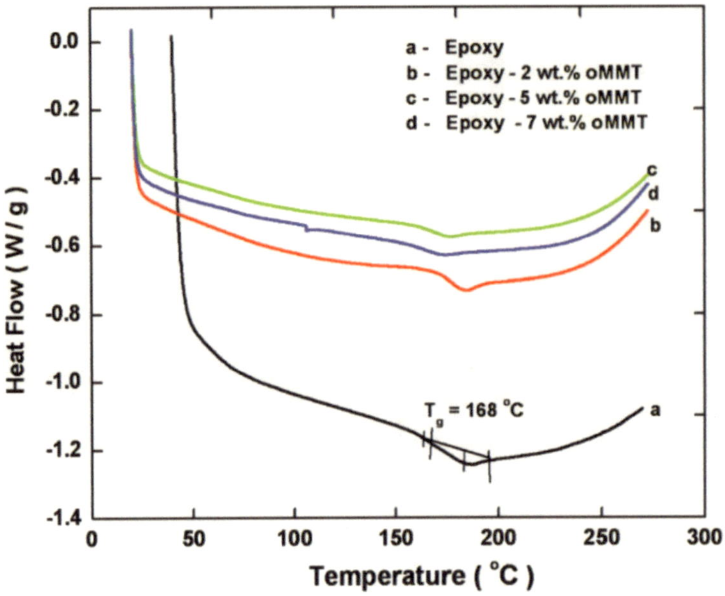

Figure 8. DSC plots of epoxy-oMMT nanocomposite.

With the incorporation of 5 and 7 wt.% of nanofiller T_g value decreases. The interfacial interaction of polymer nanocomposites and following in the formation of nanolayer in nanocomposites are concluded in the literature followed. The formation of nanolayer and its influence on T_g is also been reported. The higher value of in T_g [13, 14] when nanofiller has been added is reported in some cases whereas, some others report a smaller value of T_g due to addition of nanoparticles into base polymer matrix [15], but the majority of the proposed theories are uncertain till today. The influence of oMMT nanofiller on the curing reaction and glass transition values of the epoxy nanocomposites is investigated by DSC in order to understand the molecular mobility in the nanocomposites. When the nanofiller concentration is increased from 5 to 7 wt.%, the values of T_g in epoxy nanocomposites is observed to decrease. Similar to the results obtained in the case of epoxy filled with oMMT nanofillers in the present work, reduced value of T_g have been reported for nanoalumina filled PMMA composites [16]. The increase in T_g with the addition of 2 wt.% of nanofiller is due to few cross links that developed between the polymer matrix and nanoparticle. However, the reduction in T_g values at 5 and 7 wt.% of nanofiller loading may be due to the many reasons such as changes in molecular weight, tacticity, cross-linking density and the presence of residues from incomplete reactions. The other possible reasons could be the size of the nanofillers which are certainly larger than the free volume hole sizes (discussed in Section 2.4) in the matrix and therefore the possible slide between the chains can result in increased free volume.

Sun and co-authors [17] reported that the T_g depression is associated with the improved polymer chain dynamics because of the additional free volume at the resin-nanofiller interface. Becker and co-authors [18] remarked, interfacial interactions between polymer chains and positively or negatively charged nanofiller surface leads to the development of a polymer

http://dx.doi.org/10.5772/intechopen.78057

nanolayer near to the nanoparticle surface and this interfacial polymer nanolayer influences the T_g. These nanoparticle and polymer interactions can be developed as attractive interaction, repulsive interaction or the interaction is neutral and as such T_g can increase, decrease, or remain constant. In the present work, the results obtained for 2 wt.% nanocomposites, may be due to few cross links but for higher loading, the nanoparticle-polymer interaction found to be repulsive and therefore T_g decreases and the present results are in good concord with the published literatures [16, 17]. The cross linking density reduces due to etherification mechanism and curing agent preferably tends to attach to the surfaces and very thin layer of hardner or curing agent will surround the nanofiller particles. This thin layer of curing agent will keep the curing agent around the nanofiller particles from reacting with the epoxy resin. This 'curing agent concentration' mechanism will also cause decrease in cross-link density and thereby T_g decreases at 5 and 7 wt.% of filler loading.

3.5. Characterization of free volume content

The variation of free volume content or fractional free volume (F_v), free volume size (V_f) and longest life time (τ_3) with 2, 5 and 7 wt.% of oMMT in epoxy nanocomposites is shown in **Figure 9**.

The addition of oMMT content in epoxy resin (2-5 wt.%), causes τ_3 to increase from 1.72 to 1.75 ns, that is size or area of the free volume and free volume content increases from 7.17 to 7.45 nm^3 and 2.28 to 2.45% respectively. However, at 7 wt.% oMMT loading, both τ_3, V_f and F_v decrease slightly to 1.74 ns, 7.34 nm^3 and 2.14% respectively. The modifications has taken place may be because of addition of nanofiller in polymer matrix creates additional free volume. It further suggests that, the filler will not occupy the pre-existing free volume cavities due to

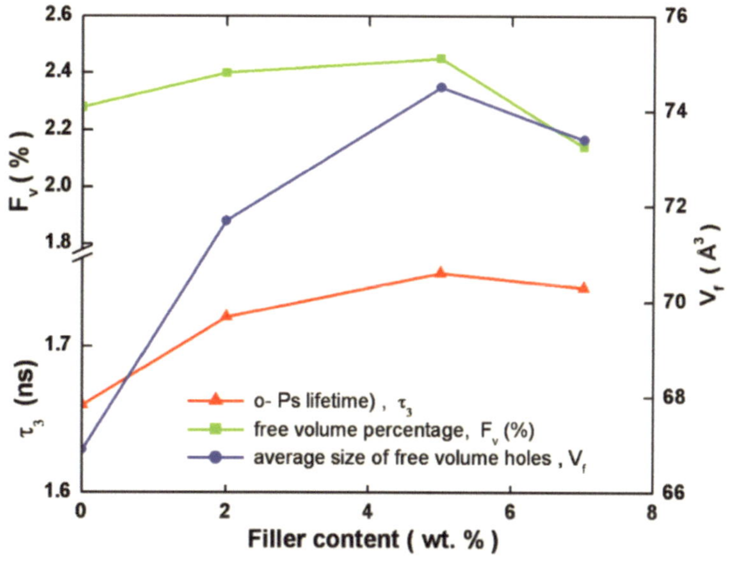

Figure 9. Plot of τ_3, V_f and F_v versus filler content (wt.%) of epoxy-oMMT nanocomposites.

bigger size of the fillers but creates additional free volume probably at the interface. The results also indicate that the layers of oMMT will result in favorable interaction with the epoxy resins and thereby the segmental motion is hindered. Hence, the significant increase in F_v percentage is not observed. The decrease in T_g is justified by the decrease in free volume content.

3.6. Effect of interface on dielectric strength (DES)

The dielectric strength (DES) is the vital properties of dielectric insulators. With reference to the discussion in Section 2.4, the properties of epoxy nanocomposites are mainly explained by interfacial interaction of polymer and of nanofillers. This interfacial area is responsible for the interaction of the electric field between the base epoxy and nanofiller. The DES of the nanocomposites depends largely on nanofiller content and even a very less quantity of nanofiller can cause improvement. When nanofiller particles are incorporated into the epoxy matrix, there is a change in morphology of the epoxy due to the interfacial interaction of epoxy with the oMMT nanofiller.

The **Figure 10** shows the variation of Dielectric strength, free volume with respect to nanofiller loading. With the addition of 2 wt.% oMMT into epoxy matrix, it is observed that T_g increases due to increase in cross linking density and hence less free charge carriers are available leading to slightly higher value of DES than that of pure epoxy. A further increase in the filler loading up to 5 wt.% of oMMT shows an increase in the DES above that of 2 wt.% of nanocomposite. In this case the T_g decreases, and hence more free charge carriers are available.

Here the effect of third interface layer also called loose layer comes into scenario. The loose polymer layers contain more traps or free volumes as discussed in Section 2.4. These charge carriers are easily and more frequently trapped in trap sites rather than in the base epoxy.

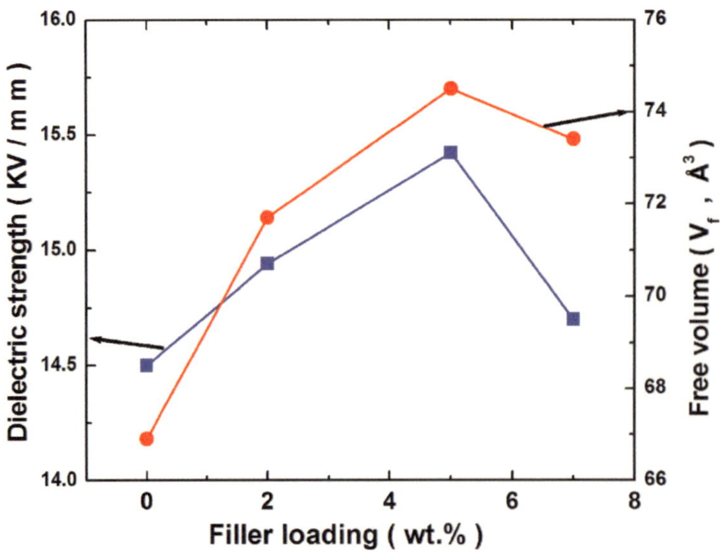

Figure 10. Plot of DES, Fv with respect to filler content (wt.%) of epoxy-oMMT nanocomposites.

Because of this cause, charge carriers are accelerated over shorter distances and have reduced mobility and kinetic energy. This process is considered as a scattering mechanism. The energy of charge carriers is distributed more evenly in the polymer and thus causes less damage in the material and prolongs the lifetime and service of the epoxy. This scattering process decreases the electric field at the electrodes and increase the voltage required for charge injection.

These fillers can also act as barriers for the penetration of the charge carriers throughout the depth of the sample. In case of 5 wt.% oMMT, the number of nanoparticles is much more and the inter particle distance is also less than 100 nm. This inter particle distance was justified by morphological studies. Hence, there is a possibility of overlapping of the loose polymer regions in the nanocomposites leading to the reduction in the loose polymer regions. This polymer layer along with a large number of fillers can also obstruct the discharge path, thereby dielectric strength increases. The increase in dielectric strength values is well supported by reduction in dielectric constant values and increase in free volume content at 2 and 5 wt.% oMMT filled epoxy-oMMT.

When oMMT filler content is increased to 7 wt.%, there is a reduction in the dielectric strength occurs due to the overlapping of the tightly bound polymer regions over the interface, since the inter particle distances are comparable to the filler diameter. The interphase region close to the nanoparticle is found to be conductive [11]. As the conductive interphase regions tends to overlap, the pure polymer region is reduced, leading to an easier conducting path for the charge transfer and thereby a reduction in the dielectric strength is observed. With the addition of 7 wt.% of oMMT into epoxy matrix, reduction in T_g has taken place, and hence more free charge carriers are available, but reduction in free volume leads to an easier conduction path for the charge carriers and thereby DES decreases. The reduction in DES at 7 wt.% of oMMT filled nanocomposite is well supported by increase in dielectric constant values and free volume content of the nanocomposites. This has been well supported by many authors. Many researchers [19] have reported that layered silicate nanofillers modify the trapping property of both isotactic and syndiotactic PP. Roy and co-authors [20] have reported that deep trap sites have been identified in SiO_2-XLPE nanocomposites through thermally stimulated current (TSC) measurements.

4. Conclusions

The following remarks are drawn:

i. Existence of hydrogen bond between nanofillers and epoxy polymer chains in the nanocomposites has been established through FTIR, T_g, and free volume measurements.

ii. The interaction takes place between epoxy and nanoparticle in small region around the oMMT nanofiller surface called as "interface region" and the nature of interaction is found to be depends on the chemical bond of the oMMT and epoxy.

iii. The interaction between the oMMT and epoxy chains has a direct effect on dielectric strength characteristics of the nanocomposite and accordingly, a three core interface model has been used to elucidate the characteristics of the interface region.

iv. The characteristics of the interface region of the nanocomposite depend on the number of oMMT nanofiller particles included in the epoxy resin and how the oMMT nanoparticle loading affects the dielectric strength.

Author details

Rashmi Aradhya[1]* and Nijagal M. Renukappa[2]

*Address all correspondence to: rash_mysore@yahoo.com

1 Department of Electrical and Electronics Engineering, Siddaganga Institute of Technology, Tumkur, India

2 Department of Electronics and Communication Engineering, Sri Jayachamarajendra College of Engineering, Mysore, India

References

[1] Tanaka T, Kozaka M, Fuse N, Ohki Y. Proposal of a multi-core model for polymer nanocomposite dielectrics. IEEE Transactions on Dielectrics and Electrical Insulation. 2005;**12**:669-681

[2] Ravikumar HB, Ranganathaiah C, Kumarswamy GN, Thomas S. Positron annihilation and differential scanning calorimetric study of poly (trimethyleneterephthalate)/EPDM blends. Polymer. 2005;**46**:2372-2380

[3] Kirkegaard P, Pedersen NJ, Eldrup M. Analysis of BiF2 poly epoxy resins. Risø National Laboratory Report. 1989. pp. 2735-2740

[4] Nakanishi H, Wang SJ, Jean YC. Microscopic surface tension studied by positron annihilation. In: Sharma SC, editor. Proceedings of the International Symposium on Positron Annihilation in Fluids. Singapore: World Scientific; 1988. pp. 292-298

[5] Tao SJ. Positronium annihilation in molecular substances. Chemical Physics. 1972;**56**: 5499-5510

[6] Eldrup M, Lightbody D, Sherwood JN. The temperature dependence of positron lifetimes in solid pivalic acid. Chemical Physics. 1981;**63**:51-58

[7] Ranganathaiah C, Kumaraswamy GN. New method of determining miscibility in binary polymer blends through hydrodynamic interaction: The free volume approach. Journal of Applied Polymer Science. 2009;**111**:577-588

[8] Raj JM, Ranganathaiah C. A new method of stabilization and characterization of the interface in binary polymer blends by irradiation: A positron annihilation study. Journal of Polymer Science Part B: Polymer Physics. 2009;**47**:619-632

http://dx.doi.org/10.5772/intechopen.78057

[9] Nelson JK, Utracki LA, Mac Crone RK, Reed CW. Role of the interface in determining the dielectric properties of nanocomposites. In: IEEE Conference on Electrical Insulation and Dielectric Phenomena (CEIDP). 2004. pp. 314-318

[10] Maity P, Kasiomayajula SV, Parameswaran V, Basu S, Gupta N. Improvement in surface degradation properties of polymer composites due to pre-processed nanometric alumina fillers. IEEE Transactions on Dielectrics and Electrical Insulation. 2008;**15**:63-73

[11] Nelson JK, Hu Y. Nanocomposite dielectric properties and implications. Journal of Physics D: Applied Physics. 2005;**38**:213-222

[12] Zou C, Fothergill JC, Rowe SW. A water-shell model for the dielectric properties of hydrated silica filled epoxy nanocomposites. In: International Conference on Solid Dielectrics (ICSD), Winchester, UK. 2007. pp. 389-392

[13] Costa LC, Devesa S, Andre P, Henry F. Microwave dielectric properties of polyethylene terephthalate (PBT) with carbon black particles. Microwave and Optical Technology Letters. 2005;**46**:61-63

[14] Roberston J, Varlow BR. The AC non liner permittivity characteristics of barium titanate filled acrylic resin. In: Proceedings of the International Conference on Properties and Applications of Dielectric Materials (ICPADM), Nagoya, Japan. 2003. pp. 761-764

[15] Singha S, Thomas MJ. Dielectric properties of epoxy Nanocomposites. IEEE Transactions on Dielectrics and Electrical Insulation. 2008;**15**:12-23

[16] Ash BJ, Schadler LS, Siegel RW. Glass-transition temperature behavior of alumina/PMMA nanocomposites. Journal of Polymer Science Part B: Polymer Physics. 2004;**42**:4371-4383

[17] Sun YY, Zhang Z, Wong CP. Rheology study of wafer level underfill. Macromolecular Materials and Engineering. 2005;**290**:1204-1212

[18] Becker O, Cheng YB, Varley RJ, Simon GP. Layered silicate nanocomposites based on various high-functionality epoxy resins: The influence of cure temperature on morphology, mechanical properties, and free volume. Macromolecules. 2003;**36**:1616-1625

[19] Ambid M, Maryz D, Teyssedre G, Lauramt C, Maontanari GC, Kaempter D, Mulhaupt R. Effect of filler concentration on dielectric behavior and on charge trapping in PP/clay nanocomposites. In: IEEE Conference on Electrical Insulation and Dielectric Phenomena (CEIDP). 2004. pp. 389-392

[20] Roy M, Nelson JK, MacCrone RK, Schadler LS. Candidate mechanisms controlling the electrical characteristics of silica/XLPE nanodielectrics. Journal of Materials Science. 2007;**42**:3789-3799

Chapter 9

Optimization of Functionally Graded Material Structures: Some Case Studies

Karam Maalawi

Additional information is available at the end of the chapter

http://dx.doi.org/10.5772/intechopen.82411

Abstract

This chapter focuses on some of the most advances made in the field of stability, dynamic, and aeroelastic optimization of functionally graded composite structures. Practical realistic optimization models using different strategies for measuring structural performance are presented and discussed. The selected design variables include the volume fractions of the composite material constituents as well as geometrical and cross-sectional parameters. The mathematical formulation is based on dimensionless quantities; therefore, the analysis can be valid for different configurations and sizes. Such normalization has led to a naturally scaled optimization model, which is favorable for most optimization techniques. Case studies include structural dynamic optimization of thin-walled beams in bending motion, optimization of drive shafts against torsional buckling and whirling, and aeroelastic optimization of subsonic aircraft wings. Other stability problems concerning fluid-structure interaction has also been addressed. Several design charts that are useful for direct determination of the optimal values of the design variables are introduced. The proposed mathematical models have succeeded in reaching the required optimum solutions, within reasonable computational time, showing significant improvements in the overall structural performance as compared with reference or known baseline designs.

Keywords: functionally graded materials, composite structures, optimum design, buckling stability, structural dynamics, fluid-structure interaction

1. Introduction

Functionally graded materials (*FGMs*) are new generation of advanced composites that have gained interest in several engineering applications such as spacecraft heat shields, high-performance structural elements, and critical engine components [1, 2]. They are formed of two or more constituent phases with a continuously variable composition producing properties that change spatially within a structure. *FGMs* possess a number of advantages that make them attractive in improving structural performance, such as maximized torsional rigidity of composite shafts [3], improved residual stress distribution and enhanced thermal properties [4], higher natural frequencies of composite beams [5], and broader aeroelastic stability boundaries of aircraft wings [6]. Actually, the concept of *FGMs* was originated in Japan in 1984 during a space project, in the form of proposed thermal barrier material capable of withstanding high-temperature gradients.

Figure 1 shows variation of the volume fraction through the thickness of a plate fabricated from ceramic and metal. Ceramic provides high-temperature resistance because of its low thermal conductivity, while metal secures the necessary strength and stiffness. *FGMs* may also be developed using fiber-reinforced layers with a volume fraction of fibers that is coordinate dependent, rather than constant, producing favorable properties or response [6]. In this chapter, much attention is given to fibrous-type and their constitutive relationships.

An excellent review paper dealing with the basic knowledge and various aspects on the use of *FGMs* and their wide applications is given by Birman and Byrd [7], who presented comprehensive discussions of the development related to stability and dynamic of *FGM* structures. Closed-form expressions for calculating the natural frequencies of an axially graded beam were derived in [8], where the modulus of elasticity was taken as a polynomial of the axial coordinate along the beam's length. An inverse problem was solved to find the stiffness and mass distributions so that the chosen polynomial serves as an exact mode shape. Qian and Batra [9] considered frequency optimization of a cantilevered plate with variable volume fraction according to simple power laws. Genetic algorithm was implemented to find the optimum values of the power exponents, which maximize the natural frequencies. They concluded that the volume fraction needs to be varied in the longitudinal direction of the plate rather than in the thickness direction. Another work presented an analytical approach for

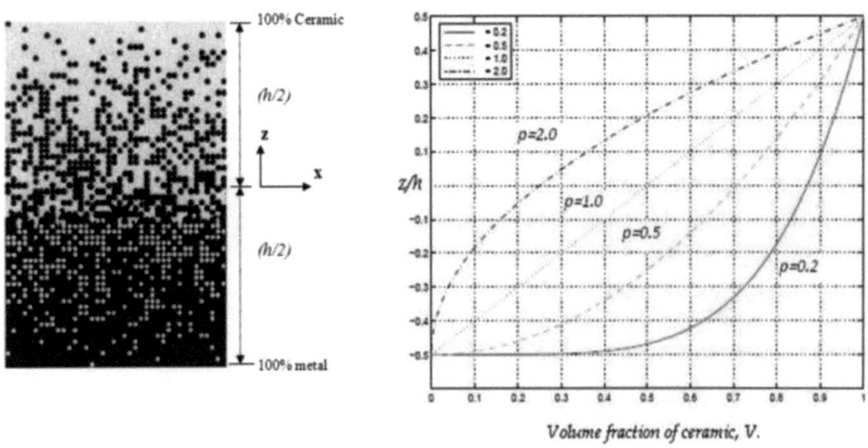

Figure 1. FGM ceramic/metal particulate composite with volume fraction graded in the vertical direction [1].

designing efficient patterns of *FGM* bars having maximized frequencies while maintaining the total mass at a constant value [10]. The distribution of the volume fractions of the material constituents was optimized using either discrete or continuous variations along the bar length.

In the context of structural stability, Elishakoff and Endres [11] considered buckling of an axially graded cantilevered column and derived closed form solution for the mode shape and the critical load. A semi-inverse method was employed to obtain the spatial distribution of the elastic modulus in the axial direction. In Ref. [12], the buckling of simply supported three-layer circular cylindrical shell under axial compressive load was analyzed. The middle layer sandwiched with two isotropic layers was made of an isotropic *FGM* whose Young's modulus varies parabolically in the thickness direction. Classical shell theory was implemented under the assumption of very small thickness/radius and very large length/radius ratios. Numerical results showed that the buckling load increases with an increase in the average value of Young's modulus of the middle layer. An exact method was given in [13] for obtaining column's designs with the maximum possible critical buckling load under equality mass constraint. Both material and wall thickness grading in the axial direction have been applied to determine the required optimal solutions. Case studies and detailed results were given for the cases of simply supported and cantilevered columns. Another work by Maalawi [14] presented a mathematical model for enhancing the buckling stability of composite, thin-walled rings/long cylinders under external pressure using radial material grading. The main structure is made of multiangle fibrous laminated layups having different volume fractions within the individual plies. This produced a piecewise grading of the material and thickness in the radial direction. The critical buckling contours are plotted for different types of materials, showing significant improvement in the overall stability limits of the structure under the imposed mass constraint.

Considering dynamic aeroelasticity of *FGM* structures, Shin-Yao [15] investigated the effect of variable fiber spacing on the supersonic flutter of composite laminates using the finite element method and quasi-steady aerodynamic theory. The formulation of the location-dependent stiffness and mass matrices due to nonhomogeneous material properties was derived. This study first demonstrates the flutter analysis of composite laminates with variable fiber spacing. Numerical results show that the sequence of the natural mode may be altered, and the two natural frequencies may be close to each other because the fiber distribution may change the distributed stiffness and mass of the plate. Therefore, it may change the flutter coalescent modes, and the flutter boundary may be increased or decreased due to the variable fiber spacing. More detailed discussions on stability, dynamic, and aeroelasticity of *FGM* structures are outlined in Ref. [16]. The attained optimal solutions were determined by applying the global search techniques [17, 18], which construct a number of starting points and use a local solver, such as *"fmincon"* routine in the MatLab optimization toolbox [19]. Global search technique is distinguished with fast converging to the global optima even if it starts with a design point far from the optimum. The local solver *"fmincon"* uses the method of sequential quadratic programming (SQP), which has a theoretical basis related to the solution of a set of nonlinear equations using Newton's method and applies Kuhn-Tucker conditions to the Lagrangian of the constrained optimization problem.

It is the main intend of this chapter to present some fundamental issues concerning design optimization of different types of functionally graded composite structures. Practical realistic optimization models using different strategies for enhancing structural dynamics, stability, as well as aeroelastic performance are presented and discussed. Case studies include frequency optimization of thin-walled box beams, optimal design of drive shafts against torsional buckling and whirling, and aeroelastic optimization of subsonic aircraft wings. Design of pipelines against flow-induced flutter and/or divergence has also been addressed. Several design charts that are useful for direct determination of the optimal values of the design variables are introduced. In all, the given mathematical models can be regarded as useful design tools, which may save designers from having to choose the values of some of their variables arbitrarily.

2. Mathematical modeling of material grading

There are different scenarios in modeling the spatial variation of material properties of a functionally graded structure. For example, Chen and Gibson [20] considered distributions represented by polynomial functions and applied Galerkin's method to calculate the required polynomial coefficients from the resulting algebraic equations. They found that the linear variation of the volume fraction is a best fit with that predicted experimentally for selected composite beam specimens. Chi and Chung [21] studied the mechanical behavior of *FGM* plates under transverse loading, where a constant Poisson's ratio and variable moduli of elasticity throughout the plate thickness were assumed. The volume fraction of the constituent materials was defined by simple power laws, and closed form solutions using Fourier series were given for the case of simply supported plates. In general, the distribution of the material properties in functionally graded structures may be designed by either continuous or piece-wise variation of the volume fraction in a specified direction. The most commonly utilized distributions are summarized in what follows.

2.1. Thickness grading

The first early model of volume fraction variation through the thickness of a plate fabricated from ceramic and metal was considered in [20]. This volume fraction is based on the mixture of metal and ceramic and is an indicator of the material composition (volumetric wise) at any given location in the thickness. If the volume fraction of ceramic is defined as "v" then the volume fraction of metal is the remainder of the material, or $(1 - v)$, assuming no voids are present. A typical example, which was considered by numerous researchers in the field [1- 4] assumed that the volume fraction "v" can be varied through the thickness coordinate z by the power law (refer to **Figure 1**):

$$v(z) = \left(\frac{1}{2}+\frac{z}{h}\right)^{p} \quad -1/2 \leq (z/h) \leq 1/2, \;\; p \geq 0 \tag{1}$$

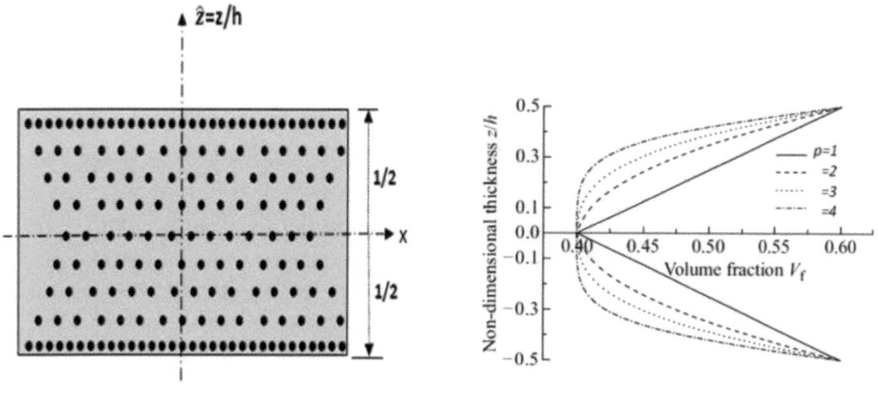

Figure 2. Thickness distribution of the fiber volume fraction in FGM beam, $v_f(0) = 40\%$, $v_f(1/2) = 60\%$ [22].

where h is the plate thickness and p is a volume fraction exponent, which dictates the amount and distribution of ceramic in the plate. With higher values of p, the plate tends toward metal while lower values tend toward ceramic ($p = 0.0$: fully ceramic, $p = \infty$: fully metal). Accordingly, the distribution of the mechanical and physical properties of FGM can be defined in terms of the material constants of the constituent phases based on a selected power-law model. Designers can vary the p-value to tailor the FGM to specific applications. In case of fibrous composites, Eq. (1) ought to be modified to account for the limits imposed on the fiber volume fractions at $\hat{z} = (zlh) = \pm 1/2$ for consideration of other strength requirements and/or manufacturing restrictions. The modified form can be expressed as follows [5]:

$$v_f(\hat{z}) = v_f(-0.5) + \left[v_f(0.5) - v_f(-0.5)\right](\hat{z} + 0.5)^P, \tag{2}$$

Another type of the power-law expression was utilized by Bedjilili et al. [22], who considered vibration of fibrous composite beams with a variable volume fraction through the thickness of the cross section, as shown in **Figure 2**. It was concluded that by varying the fiber volume fraction within the beam thickness to create a FGM, certain vibration characteristics are significantly affected. The utilized formula was given as:

$$\begin{gathered} v_f(\hat{z}) = v_f(0) + \left[v_f(0.5) - v_f(0)\right](2|\hat{z}|)^P, \\ -0.5 \le \hat{z}\left(=\frac{z}{h}\right) \le 0.5, \quad p \ge 0 \end{gathered} \tag{3}$$

2.2. Spanwise grading

Some researchers considered grading of the fiber volume fraction in the spanwise (longitudinal) direction of a composite plate. Librescu and Maalawi [6] investigated optimization of composite wings using the concept of material grading in the spanwise direction. Both continuous and discrete distributions of the fiber volume fractions were considered in the developed optimization models. The following power-law expression was implemented:

$$v_f(\hat{y}) = v_{fr}\left(1 - \beta_f \hat{y}^P\right), \quad 0 \le \hat{y}\left(= \frac{y}{L}\right) \le 1$$
$$\beta_f = (1 - \Delta_f), \Delta_f = v_{ft}/v_{fr} \tag{4}$$

where v_{ft} and v_{fr} are the fiber volume fractions at wing tip and root, respectively. Δ_f is called the tapering ratio of the fiber volume fraction. **Figure 3** shows the different patterns of the fiber volume fraction distribution for different values of the power exponent p. Both configurations of fibers aligned in the transverse (chordwise) and in the longitudinal (spanwise) directions are shown. The volume fraction is constrained to lie between 25% and 75% in order not to violate other strength and manufacturing requirements.

A more general distribution, given in Eq. (5), was tried by Shih-Yao [15], who applied it successfully to investigate the effect of grading on the supersonic flutter of rectangular composite plates.

$$v_f(\hat{y}) = v_{fr}\left[\beta_f (1 - \hat{y}^n)^p + \Delta_f\right], n = 1, 2, 3 \quad p \ge 0 \tag{5}$$

2.3. Determination of mechanical properties

A variety of approaches have been developed to predict the mechanical properties of fibrous composite materials [23]. The common approaches fall into the following general categories: mechanics of materials; numerical methods; variational approach; semiempirical formulas; experimental measurements. Mechanics of materials approach is based on simplifying assumptions of either uniform strain or uniform stress in the constituents. Its predictions can be adequate only for longitudinal properties of unidirectional continuous fibrous composites. Numerical methods using finite difference, finite element, or boundary element methods yield the best predictions; however, they are time-consuming and do not yield closed-form expressions. Variational methods based on energy principles have been developed to establish bounds (inequality relations) on the effective properties. The bounds are close to each other in the case of longitudinal properties, but they can be far apart in the case of transverse and shear properties. Semiempirical relationships have been developed to avoid the difficulties with the above theoretical approaches and to facilitate computations. The so-called Halpin-Tsai relationships have consistent forms for all properties of fibrous composite materials and can be used to predict the effects of a large number of system variables. **Table 1** summarizes the mathematical formulas for determining the equivalent mechanical and physical properties for known type and volume fractions of the fiber (V_f) and matrix (V_m) materials [23]. The 1 and 2 subscripts denote the principal directions of an orthotropic lamina, defined as follows: direction (1) principal fiber direction, also called fiber longitudinal direction; direction (2) in-plane direction perpendicular to fibers, transversal direction. The factor ξ is called the reinforcing efficiency and can be determined experimentally for specified types of fiber and matrix materials. Whitney [24]

http://dx.doi.org/10.5772/intechopen.82411

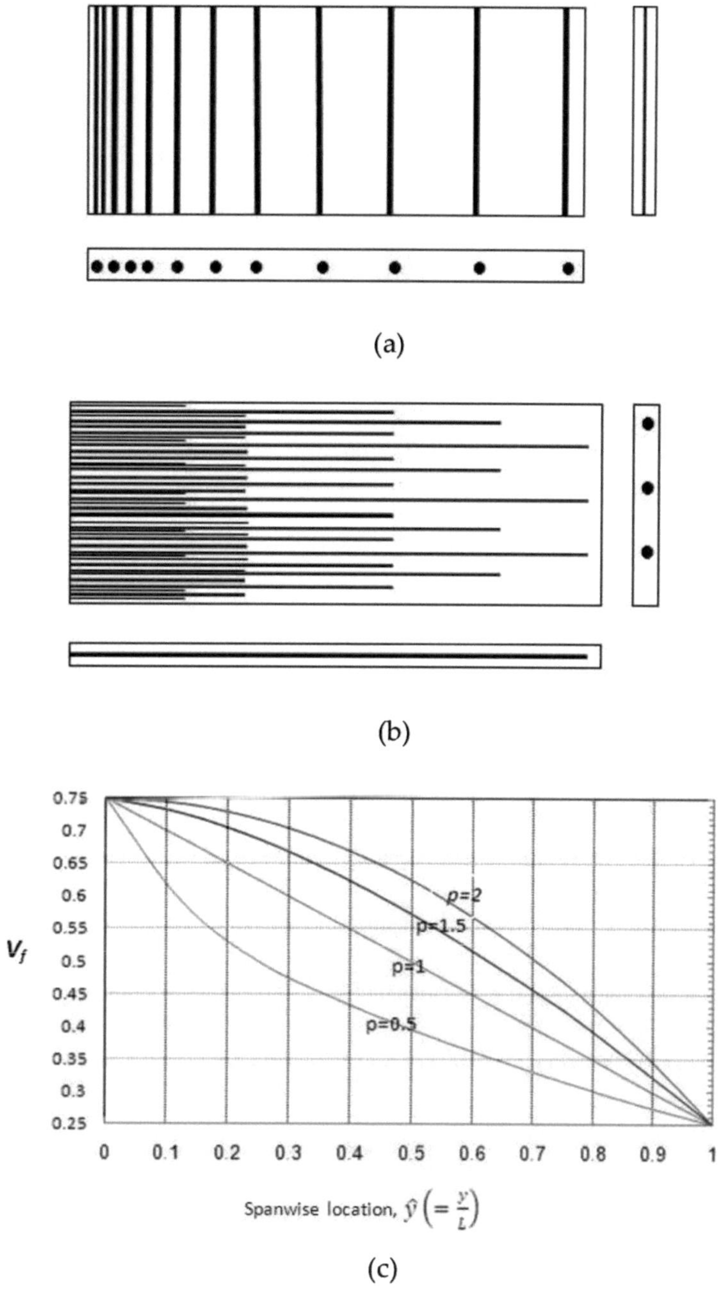

Figure 3. Spanwise grading of fibers in a fibrous composite plate [6]. (a) Fibers aligned in chordwise direction, (b) fibers aligned in spanwise direction, and (c) fiber volume fraction distribution for different power exponents.

Property	Mathematical formula*
Young's modulus in direction (1) E_{11}	$E_m V_m + E_{1f} V_f$
Young's modulus in direction (2) E_{22}	$E_m (1 + \xi\eta V_f)/(1-\eta V_f)$; $\eta = (E_{2f} - E_m)/(E_{2f} + \xi E_m)$
Shear modulus G_{12}	$G_m (1 + \xi\eta V_f)/(1-\eta V_f)$; $\eta = (G_{12f} - G_m)/(G_{12f} + \xi G_m)$
Poisson's ratio ϑ_{12}	$\vartheta_m V_m + \vartheta_{12f} V_f$
Mass density ρ	$\rho_m V_m + \rho_f V_f$

*Subscripts "*m*" and "*f*" refer to properties of matrix and fiber materials, respectively. Assuming no voids are present, then $V_m + V_f = 1$.

Table 1. Halpin-Tsai semiempirical relations for calculating composite properties [23].

suggested the range $1 < \xi < 2$ depending on the fiber array type, for example, hexagonal, square, etc. Usually, ξ is taken equal to *1.0* for theoretical analysis procedures in the case of carbon or glass fibrous composite laminates.

3. Frequency optimization of *FGM* thin-walled box beams

This section presents a mathematical model for optimizing the dynamic performance of thin-walled *FGM* box beams with closed cross sections. The objective function is to maximize the natural frequencies and place them at their target values to avoid the occurrence of large amplitudes of vibration. The variables considered include fiber volume fraction, fiber orientation angle, and ply thickness distributions. Various power-law expressions describing the distribution of the fiber volume fraction have been implemented, where the power exponent was taken as the main optimization variable [25]. The mass of the beam is kept equal to that of a known reference beam. Side constraints are also imposed on the design variables in order to avoid having unacceptable optimal solutions. A case study on the optimization of a cantilevered, single-cell spar beam made of carbon/epoxy composite is considered. The results for the basic case of uncoupled bending motion are given.

3.1. Structural dynamic analysis

Figure 4 shows a slender, composite thin-walled beam constructed from uniform segments, each of which has different cross-sectional dimensions, material properties, and length. Tapered shapes of an actual blade or wing spar can be adequately approximated by such a piecewise structural model with a sufficient number of segments. The various parameters and variables are normalized with respect to a reference beam, which is constructed from just one segment with single unidirectional lamina having equal fiber and matrix volume fractions, that is, $V_{fo} = V_{mo} = 50\%$. The different quantities are defined in the following:

N_s = number of segments (panels).

j = subscript for the j-th segment, $j = 1, 2, \ldots\ldots Ns$.

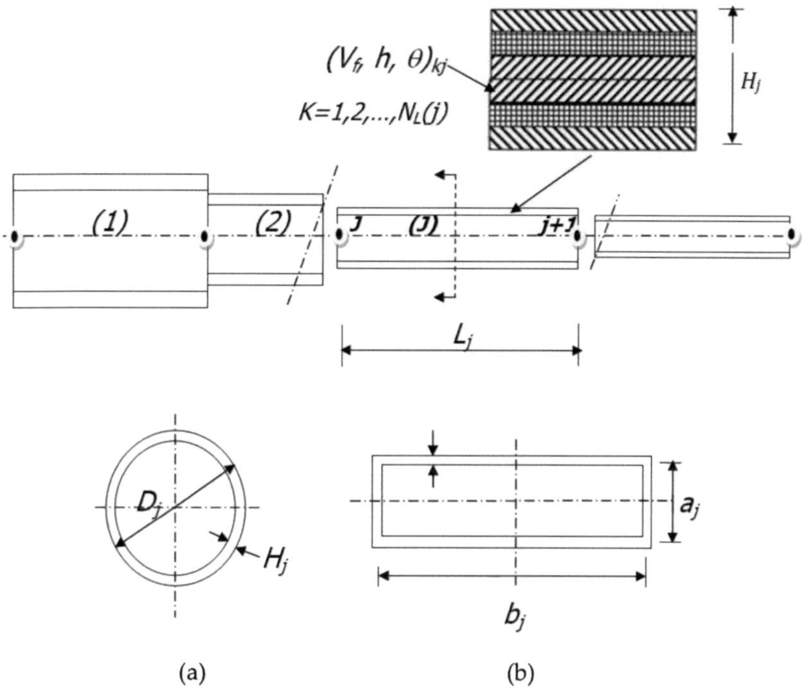

Figure 4. General configuration of a multisegment, composite box beam [25]. (a) Circular cross section, (b) rectangular cross section.

$N_L(j)$ = number of layers in the j-th segment.

k = subscript for the k-th layer, $k = 1, 2, \ldots N_L(j)$.

$\hat{L}_j = (L_j/L_o)$ = normalized length of the j-th segment.

$\hat{L} = (L/L_o) = \sum_{j=1}^{N_s} \hat{L}_j$ = normalized total beam length.

$\hat{H}_j = (H_j/H_0) = \hat{H}_j = \sum_{k=1}^{N_L(j)} \hat{h}_{kj}$= normalized total wall thickness of the j-th segment.

$\hat{h}_{kj} = (h_{kj}/H_0)$ = normalized thickness of the k-th layer in the j-th segment.

θ_{kj} = fiber orientation angle in the k-th layer in the j-th segment.

$\hat{\Gamma}_j = \Gamma_j/\Gamma_o = \oint_j ds/\oint_o ds$ = normalized circumference of the j-th segment cross section.

$\Gamma_j = \pi D_j$ for circular C.S., $\Gamma_j = 2(a_j + b_j)$ for rectangular C.S.

$\hat{\rho}_{kj} = \rho_{kj}/\rho_o$ = normalized density of the k-th layer in the j-th segment.

$V_{f,kj}$ = fiber volume fraction in the k-th layer in the j-th segment.

$\rho_{kj} = \rho_f V_{f,kj} + \rho_m (1 - V_{f,kj})$, $\rho_o = 0.5(\rho_f + \rho_m)$.

ρ_f = fiber mass density, ρ_m = matrix density.

$\hat{m}_j = (m_j/m_o)$ = normalized mass per unit length of the j-th segment.

$m_j = \Gamma_j \sum_{k=1}^{N_L(j)} \rho_{kj} h_{kj}$ = mass per unit length of the j-th segment, $m_o = \Gamma_o \rho_o H_o$.

I_j = mass polar moment of inertia per unit length of the j-th segment.

$= \sum_k^{N_L(j)} \oint \rho_{kj} h_{kj} (y^2 + z^2)\, ds.$

The normalized total structural mass is given by the expression:

$$\widehat{M}_s = M_s/M_0 = \sum_{j=1}^{N_s} \widehat{M}_j = \sum_{j=1}^{N_s} \hat{m}_j \hat{L}_j = \sum_{j=1}^{N_s} \hat{\Gamma}_j \hat{L}_j \sum_{k=1}^{N_L(j)} \hat{\rho}_{kj} \hat{h}_{kj} \tag{6}$$

where $M_o = m_o L_o = \Gamma_o \rho_o H_o L_o$ is the total mass of the uniform baseline design. A quantity with subscript "o" refers to a reference beam parameter.

3.1.1. Constitutive relationships

The displacement field of anisotropic thin-walled closed cross-sectional beams was derived by Dancila and Armanios [26], who used a variational asymptotic approach to obtain the following constitutive equations:

$$\begin{Bmatrix} F_x \\ M_x \\ M_y \\ M_z \end{Bmatrix} = \begin{bmatrix} C_{11} & C_{12} & C_{13} & C_{14} \\ C_{12} & C_{22} & C_{23} & C_{24} \\ C_{13} & C_{23} & C_{33} & C_{34} \\ C_{14} & C_{24} & C_{34} & C_{44} \end{bmatrix} \begin{Bmatrix} U_1' \\ \varphi' \\ U_3'' \\ U_2'' \end{Bmatrix} \tag{7}$$

where F_x, M_x, M_y, and M_z stand for the axial force, torsional, and bending moments, respectively, and C_{mn} are the cross-sectional stiffness coefficients derived in terms of closed-form integrals of the geometry and material constants. The notations U_1, U_2, U_3, and ϕ are the kinematic variables representing the average displacements and rotation of the cross section. The primes denote differentiation with respect to x.

3.1.2. Equations of motion

The general equations of motion for the free vibration analysis are derived using Hamilton's principle and expressed in terms of the kinematic variables, where it was shown that a closed form solution is not available [25]. However, particular choices of cross-sectional shape and layup can produce zero coupling coefficients in the equations of motion. Two special layup configurations can be considered, namely circumferentially uniform stiffness (***CUS***) and circumferentially asymmetric stiffness (***CAS***). The equations of the *CUS* type consist of two coupled equations for extension-twist and two uncoupled bending equations. For the *CAS* type, the extension displacement (U_1) is uncoupled, as well as the edgewise bending (U_2), while the flapping displacement (U_3) is coupled with twist (φ). The general solution can be obtained by separating the space and time variables, where the time dependence is assumed to be harmonic with circular frequency, ω. The solutions for the uncoupled axial and bending equations are straightforward, while those for the coupled equations involve much mathematics [26].

3.1.3. Solution procedure of uncoupled bending motion

The basic important case to be considered first is the uncoupled bending response, which exists in both *CUS* and *CAS* layup configurations. Using the multisegment model depicted in **Figure 4** and considering flapping motion (U_3), the associated eigenvalue problem can be written directly in the form:

$$C_{33,j}U_3'''' - \omega^2 m_j U_3 = 0 \tag{8}$$

which must be satisfied over the length L_j of any segment composing the beam structure. Normalizing with respect to the reference beam, we get:

$$\hat{U}_3'''' - \hat{\beta}_j^4 \hat{U}_3 = 0 \tag{9}$$

where $\hat{\beta}_j = \sqrt{\hat{\omega}}\left(\hat{m}_j/\hat{C}_{33,j}\right)^{1/4}$, $\hat{C}_{33,j} = C_{33,j}/C_{33,0}$, and $\hat{\omega} = \omega L_0^2\left(\frac{m_0}{C_{33,0}}\right)^{1/2}$. Eq. (9) must be satisfied in the interval $0 \le \overline{x} \le \hat{L}_j$, where $\overline{x} = \hat{x} - \hat{x}_j$ is a local coordinate of the *j*-th segment and $\hat{x} = (x/L_o)$. The general solution is well known to be:

$$\hat{U}_3(\overline{x}) = a_1 \sin\hat{\beta}_j\overline{x} + a_2 \cos\hat{\beta}_j\overline{x} + a_3 \sinh\hat{\beta}_j\overline{x} + a_4 \cosh\hat{\beta}_j\overline{x} \tag{10}$$

Expressing the constants a_i, $i = 1, 2, 3, 4$ in terms of the state variables vector $\{S\}^T$ = $\{U_3 \quad -U_3' \quad -C_{33}U_3'' \quad -C_{33}U_3'''\}^T$ at both ends of the *j*-th segment, we get

$$\{S\}_{j+1} = \left[T^{(j)}\right]\{S\}_j \tag{11}$$

where $[T^{(j)}]$ is called the transfer matrix of the *j*-th segment with its elements given in detail in Ref. [25]. The state variable vectors can be computed progressively along the length of the beam by applying continuity among the interconnecting joints of the different segments composing the beam structure. An overall transfer matrix denoted by *[T]*, which relates the state variables at both ends of the beam, can be obtained from the following matrix multiplication:

$$[T] = \left[T^{(Ns)}\right]\left[T^{(Ns-1)}\right]\ldots\ldots\left[T^{(2)}\right]\left[T^{(1)}\right] \tag{12}$$

The required frequency equation for determining the natural frequencies can then be obtained by applying the associated boundary conditions and considering only the nontrivial solution of the resulting matrix equation.

3.2. Formulation of the optimization problem

Several design objectives can exist in structural optimization including minimum mass, maximum natural frequencies, minimum manufacturing cost, etc. [17]. Considering the reduction of vibration level, two optimization alternatives can be formulated, namely, frequency placement by separating the natural frequencies from the harmonics of the excitations or direct maximization of the natural frequencies. The latter can ensure a simultaneous balanced improvement in both stiffness and mass distributions of the vibrating structure. The related optimization problems are usually formulated as nonlinear mathematical programming problem where iterative techniques are implemented for finding the optimal solution in the selected design space. Numerous computer programs [18] are available to solve nonlinear optimization models, which can be interacted with structural and eigenvalue analyses routines. The *MATLAB* toolbox optimization routines can be useful in solving some types of unconstrained and constrained optimization problems. One of the most commonly applied routines that find the constrained optima of a nonlinear merit function of many variables is named *"fmincon"* [19].

3.2.1. Basic optimization problem

Before performing the necessary mathematics, it is essential to recognize that design optimization is only as meaningful as its core model of structural analysis. Any deficiencies therein will absolutely be affected in the optimization process. Consider the basic problem of a uniform cantilevered, thin-walled, single-cell spar constructed from just one segment with one unidirectional lamina ($Ns = 1$, $N_L = 1$). The total length and outer cross-sectional dimensions are given preassigned values equal to those of the baseline design. The remaining set of variables is, therefore, $\underline{X} = \left(V_f, \hat{H}, \theta\right)$. The associated frequency equation for such a basic case is:

$$\cos\hat{\beta}\hat{L}\,\cosh\hat{\beta}\hat{L} = -1, or\ \cos\sqrt{\hat{\omega}}\left(\hat{m}/\hat{C}_{33}\right)^{\frac{1}{4}}\hat{L}\,\cosh\sqrt{\hat{\omega}}\left(\hat{m}/\hat{C}_{33}\right)^{\frac{1}{4}}\hat{L} = -1 \tag{13}$$

It is seen that $\sqrt{\hat{\omega}}$ is an implicit function of the design variables and can be calculated numerically by any suitable method such as Newton-Raphson or the Bisection method. However, the frequency equation can be solved directly for the whole term $\sqrt{\hat{\omega}}\left(\hat{m}/\hat{C}_{33}\right)^{\frac{1}{4}}\hat{L}$ without regard to the specific values of the design variables. The computed roots are:

$$\sqrt{\hat{\omega}_i} = \left(\hat{C}_{33}/\hat{m}\right)^{\frac{1}{4}}\left(\frac{1}{\hat{L}}\right)(1.8751, 4.6941, 7.8548, \ldots\pi(i-0.5)\quad i\geq 4 \tag{14}$$

In Eq. (14), the frequency parameter $\sqrt{\hat{\omega}_i}$ can be imagined as an explicit function of the design variables. So, for prescribed values of the design variables within the domain of side constraints, $\sqrt{\hat{\omega}_i}$ can be obtained directly from the above equation. Therefore, it is possible to place the frequency at its desired value and obtain the corresponding value of any one of the design variables directly from Eq. (14). The selected composite material of construction is made of epoxy-*3501-6* and carbon-*AS4* (see **Table 2**), which has favorable properties and is highly recommended in many applications of civil, aerospace, and mechanical engineering [23].

Figure 5 depicts the functional behavior of the dimensionless fundamental frequency parameter $\sqrt{\hat{\omega}_1}$ combined with the structural mass constraint ($\hat{M}_s = 1$). The imposed side constraints are:

$$(0.25, 0.75, -\pi/2) \leq \left(V_f, \hat{H}, \theta\right) \leq (0.75, 1.25, \pi/2\,) \tag{15}$$

It is remarked that the function is continuous and well behaved everywhere in ($V_f - \theta$) design space. The contours are symmetric about the horizontal line $\theta = 0$ where the constrained global maxima occurs when the fiber volume fraction reaches its upper limiting value. It can then be concluded that the unidirectional lamina is favorable when considering beam designs with

Property	Fiber: Carbon-AS4	Matrix: Epoxy-3501-6
Density (g/cm^3)	$\rho_f = 1.81$	$\rho_m = 1.27$
Modulus of elasticity (GPa)	$E_{1f} = 235.0$ $E_{2f} = 15.0$	$E_m = 4.30$
Modulus of rigidity (GPa)	$G_{12f} = 27.0$	$G_m = 1.60$
Poisson's ratio	$\nu_{12f} = 0.20$	$\nu_m = 0.35$

Table 2. Material properties of fiber and matrix materials [23].

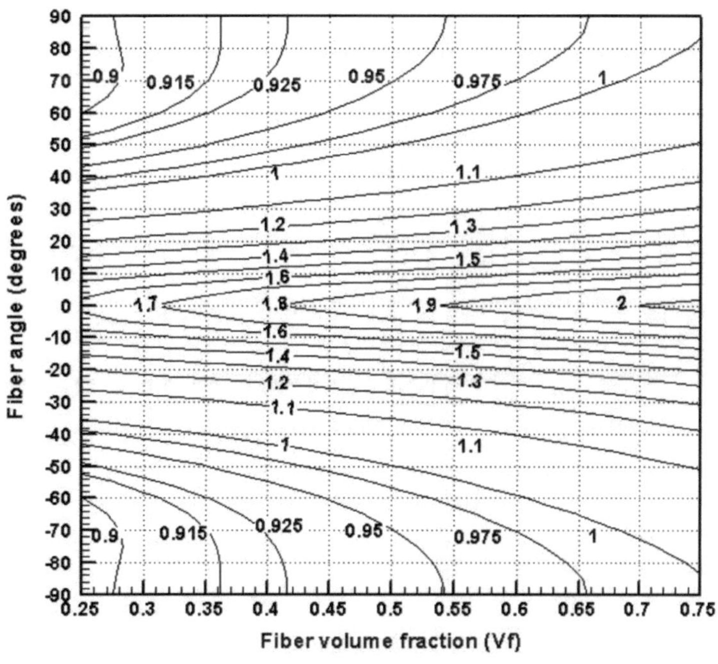

Figure 5. Level curves of $\sqrt{\hat{\omega}_1}$ function augmented with the constraint $\hat{M}_s = 1$ in $(V_f - \theta)$ design space ($Ns = 1, N_L = 1, \hat{L} = 1$).

maximum bending frequency. The optimal design point was found to be $(V_f, \hat{H}, \theta) = (0.75, 0.92, 0)$ at which $\left(\sqrt{\omega_1}\right)_{max} = 2.02589$. This corresponds to an optimization gain of about 8.04% as measured from the reference value *1.8751*. Before ending this section, it is interesting to address here the dual optimization problem of minimizing the total structural mass under preserved frequency ($\sqrt{\hat{\omega}_1}$=1.8751). The optimal solution was calculated to be $(V_f, \hat{H}, \theta) = (0.50, 0.915, 0)$ and $\hat{M}_{s,min} = 0.915$, which corresponds to a mass saving of *8.5%* as compared to the baseline design.

A couple of words are stated here regarding the side constraints in Eq. (15). First of all, it is reminded that the main focus of the present study is to optimize the fiber volume fraction in order to achieve higher values of the natural frequencies without mass penalty. The optimization is performed with respect to a known baseline design, which is considered to be conservative having reserve strength to withstand severe dynamic loads. The imposed side constraint on the total wall thickness, normalized with respect to that of the baseline design, is included for consideration of strength and stability requirements, which are not considered in the present study. So, the imposed limits with a percentage of 25% below or above that of the baseline can be practically accepted for the given model formulation. On the other hand, appropriate values of the upper and lower bounds imposed on the fiber volume fraction are chosen to avoid having unacceptable designs from the manufacture point of view. For example, the filament winding is usually associated with the highest fiber volume fractions. With careful control of fiber tension and resin content, values of around 75% would be reasonable [27].

3.2.2. *Optimization model for discrete grading*

A comprehensive analysis and formulation of discrete optimization models for beam structures considering both stability and dynamic performance were formulated in [28], where mathematical programming coupled with finite element analysis procedures was implemented. For the case of a two-segment spar beam, ($Ns = 2$, $N_L = 1$), the reduced optimization problem can be defined as follows:

$$\begin{aligned} &Minimize\ F(\underline{X}) = -\left(\sqrt{\hat{\omega}_1}\right) \\ &Subject\ to\ \hat{M}_s = 1 \\ &\sum_{1}^{Ns} \hat{L}_j = 1 \\ &(0.25, 0.0) \le \left(Vfj, \hat{L}_j \right)_{j=1,2} \le (0.75, 1.0) \end{aligned} \tag{16}$$

Using the equality constraints, two of the design variables can be expressed in terms of the other two variables. **Figure 6** shows the functional behavior of the dimensionless frequency combined with the structural mass constraint. It is remarked that the function is well defined in the feasible domain of the selected design space $(V_A - \hat{L})_1$. Two empty regions can be observed at the upper left and right parts of the design space, where violation of the equality

http://dx.doi.org/10.5772/intechopen.82411

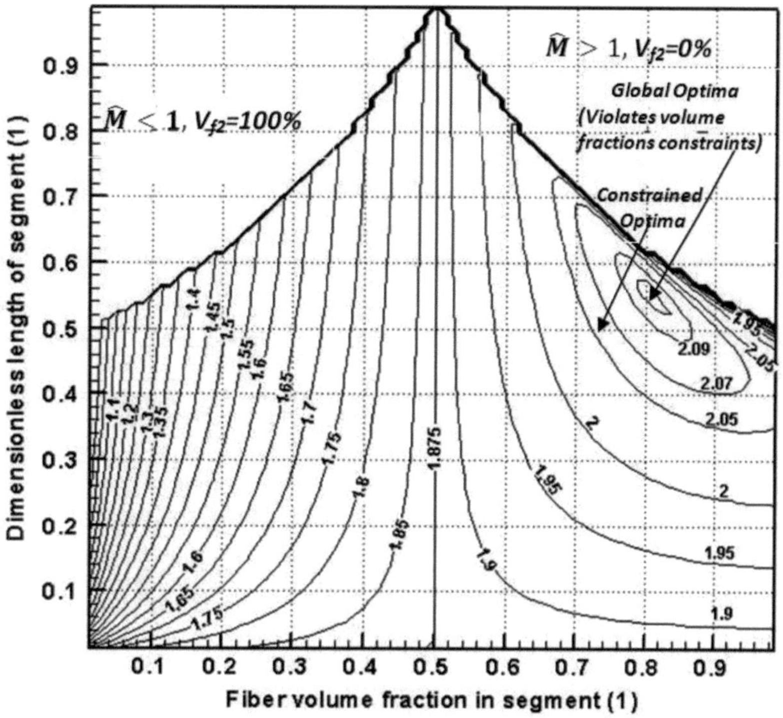

Figure 6. Level curves of $\sqrt{\hat{\omega}_1}$ function augmented with $\hat{M}_s = 1$ in $\left(V_{f1}, \hat{L}_1\right)$ design space (*Ns* = 2).

mass constraint is indicated. In the left one, the fiber volume fraction is equal to *100%*, violating the imposed side constraint. The feasible domain is seen to be split into two distinct zones separated by the baseline contour, which is represented by the vertical line V_{f1} = *50%*. The constrained optimum is to found to be $(V_{fj}, \hat{L}_j)_{j=1,2}$ = (0.75, 0.50), (0.25, 0.50) corresponding to $(\sqrt{\hat{\omega}_1})_{max}$ = 2.0645 with 10.10% optimization gain.

3.2.3. *Optimization model for continuous grading*

For continuous grading models, the associated optimization problem is cast as follows: find the design variables vector $\underline{X} = (\Delta_f, p)$, which minimizes the objective function:

$$F(\underline{X}) = -\sqrt{\hat{\omega}_1}$$

subject to the constraints:

$$\begin{gathered} \hat{M}_s = 1 \\ 0.33 \le \Delta_f \le 3.0 \\ P \ge 0 \end{gathered} \qquad (17)$$

Solutions obtained by applying the power-law model of Eq. (3) have shown that no improvements can be achieved using grading of the fiber volume fraction in the thickness direction. On the other hand, grading in spanwise direction has shown some interesting results. Considering spanwise grading according to Eq. (4), **Figure 7** depicts the level curves of the fundamental frequency parameter $\sqrt{\hat{\omega}_1}$ combined with the mass constraint in the design space (Δ_f, p). It is observed that the feasible domain is bounded from below and above by the constraint curves corresponding to the upper and lower bounds imposed on the fiber volume fractions at tip and root. The horizontal line $\Delta_f = 1.0$ (i.e., $V_f = 50\%$ at root and tip) split the domain into two zones. The lower zone encompasses the constrained optimum solution: $(\sqrt{\hat{\omega}_1})_{max}$ = *2.01875* at the design point $(\Delta_f, P)_{opt.}$ = *(0.34, 1.01)*.

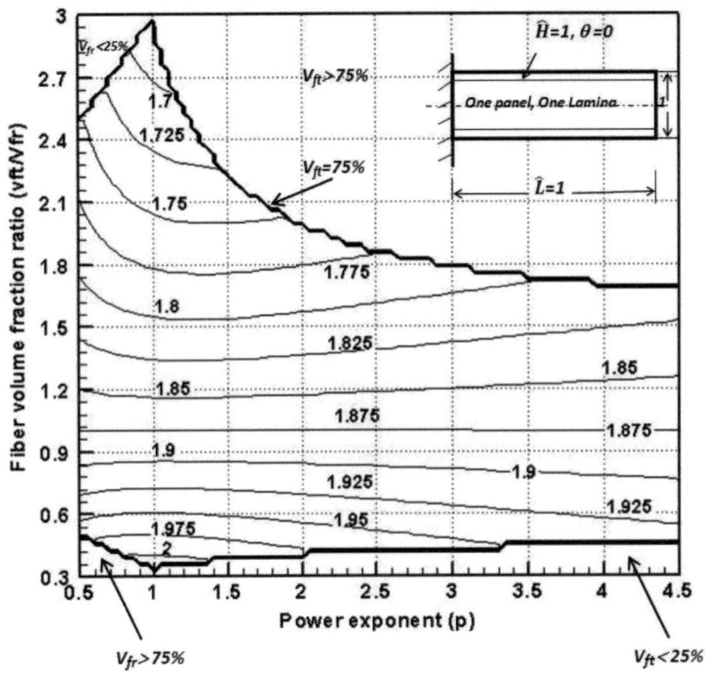

Figure 7. Level curves of $\sqrt{\hat{\omega}_1}$ function augmented with $\hat{M}_s = 1$ in (Δ_f, p) design space ($Ns = N_L = 1$) with spanwise grading *"Eq. 4."*

V_f—power-law model	$(\Delta_f, p)_{opt.}$,	$(\sqrt{\omega_1})_{max}$,	*gain %*
Thickness grading (*Eq. (3)*)	*(1.0, 0.0),*	1.8751	0.0%
Spanwise grading (*Eq. (4)*)	*(0.34, 1.01),*	2.01875,	7.66%
Spanwise grading (*Eq.* 5)			
$n = 1$	(0.34, 1.02),	2.01938,	7.70%
$n = 2$	(0.34, 2.425),	2.04813,	9.23%
$n = 3$	(0.34, 5.175),	2.06125,	9.93%

Table 3. Optimal solutions using different grading patterns ($\hat{M}_s$ = 1).

Table 3 summarizes the attained optimal solutions for the different grading patterns. It is seen that the highest optimization gain is obtained by using spanwise grading of Eq. (5) with the coordinate exponent $n = 3$.

4. Optimization of FGM drive shafts against torsional buckling and whirling

One of the important design issues in mechanical industries is the buckling and whirling instabilities that may arise from the loads applied to a power transmission shaft. These instabilities result in a reduced control of the vehicle, undesirable performance, and often cause damage, sometimes catastrophic, to the vehicle structure. Therefore, by incorporating such considerations into an early design optimization [29], the design space of a power transmission shaft will be reduced such that undesirable instability effects can be avoided during the range of the vehicle's mission profile. **Figure 8** shows an idealized structural model of a long, slender composite shaft having circular thin-walled cross section. The main structure is constructed solely of functionally graded, fibrous composite materials. The laminate coordinates are defined by x parallel to the shaft axis, y points to the tangential direction, and z points to the radial direction. Predictions of both torsional buckling and whirling instabilities are based on simplified analytical solutions of equivalent beam and shell structures. The coupling between bending and torsional deformations, introduced by the composite construction, and its influence on such instabilities is considered.

4.1. Torsional buckling optimization problem

Bert and Kim [30] derived the governing differential equations of torsional buckling in the form:

$$\begin{gathered} N_{x,x} + N_{yx,y} - 2Tu_{,xy} = 0 \\ N_{xy,x} + N_{y,y} + \left(M_{xy,x}/R\right) + \left(M_{y,y}/R\right) - 2T\left(v_{,y} + w_{,x}/R\right) = 0 \\ M_{x,xx} + \left(M_{xy} + M_{yx}\right)_{,xy} + M_{y,yy} - N_y/R + 2T\left(v_{,x}/R - w_{,xy}\right) = 0 \end{gathered} \tag{18}$$

where N_x and N_y are the normal forces, N_{xy} and N_{yx} are shear forces, M_x and M_y are bending moments, and M_{xy} and M_{yx} are torsional moments. All are applied to the midsurface and measured per unit wall thickness of the shaft. T is the applied torque, R is the mean radius, and (u, v, w) are the displacements of a generic point on the middle surface of the shaft wall. An iterative process is outlined in Ref. [30] for calculating the buckling torque for specified boundary conditions. There are other simple empirical equations based on experimental studies that can give a reasonable estimate of the buckling torque. The most commonly used formula for the case of simply supported shaft is [31]:

$$T_{cr} = \left(2\pi R^2 H\right)(0.272)(E_x)^{0.25}\left(E_y\right)^{0.75}(H/R)^{1.5} \tag{19}$$

where T_{cr} is the critical buckling torque and H is the total wall thickness of the shaft. Expressions of the equivalent modulii of elasticity in the axial (E_x) and hoop (E_y) directions for

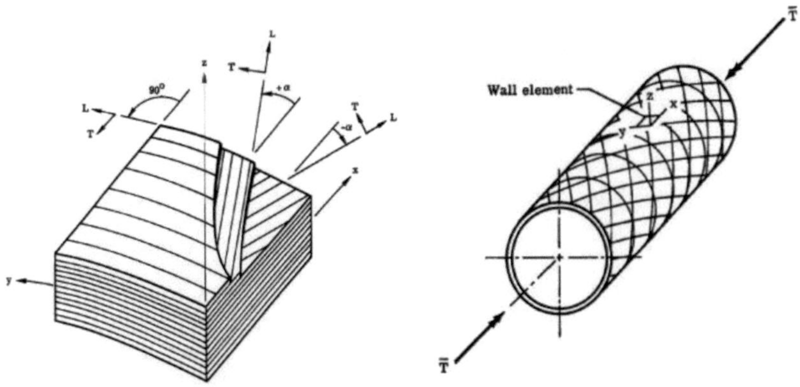

Figure 8. Shaft model and definition of reference axes.

symmetric and balanced laminates are given in Ref. [31]. The various parameters and variables are normalized with respect to known baseline design, which is constructed from cross-ply laminates $[0^o/90^0]_N$ with equal volume fractions of the fibers and matrix materials, that is, $V_f = V_m = 50\%$. Optimized shaft designs shall have the same transmitted power, length, outer diameter, boundary conditions, and material properties of those known for the baseline design. The different dimensionless quantities are defined in Ref. [31]. The optimal torsional buckling problem is to find the design variables vector $\vec{X} = \left(V_f, \theta, \hat{h}\right)_{k=1,2,\ldots N_L}$, which minimizes the objective function:

$$\begin{array}{lll} & \text{Minimize} & F = -\hat{T}_{cr} \\ \text{subject to} & \text{mass limitation}: & \hat{M} - 1 \le 0 \end{array} \tag{20}$$

$$\begin{array}{ll} \text{Torsional strength}: & \left(\dfrac{\tau_{max}}{\tau_{allow}}\right) - 1.0 \le 0 \\ \text{Whirling}: & \hat{\Omega}_{max} - \hat{\Omega}_{cr} \le 0 \end{array} \tag{21}$$

$$\begin{array}{ll} \text{Side constraints}: & \left(0.30, -\dfrac{\pi}{2}, 0.015\right) \le \left(V_f, \theta, \hat{h}\right)_{k=1,2,\ldots N_L} \le \left(0.70, \dfrac{\pi}{2}, 0.20\right) \\ & 0.75 \le \displaystyle\sum_{k=1}^{N_L} \hat{h}_k \le 1.25 \end{array} \tag{22}$$

where $\hat{T}_{cr} = T_{cr}/T_{cro}$, $\hat{\Omega} = (\Omega * 2\pi)/(60\omega_{1,o})$ are the dimensionless critical torque and rotational speed, respectively. The baseline design parameters are denoted by subscript "o." $\tau_{max} = \left(T_{max}/2\pi R^2 H\right)$ is the maximum shear stress, T_{max} is the maximum applied torque, and τ_{allow} is the allowable shear stress that can be calculated according to the embedded material properties and volume fraction of the fiber [23]. This optimization problem may be thought as a search in an $(3N_L)$ dimensional space for a point corresponding to the minimum value of the objective function and such that it lies within the region bounded by subspaces representing the constraint functions. It must be noted that the outside dimensions (outer diameter and length) of

http://dx.doi.org/10.5772/intechopen.82411

the shaft are restricted by the available interior space of the vehicle and will be considered as preassigned parameters in the present model formulation. The first case study to be examined herein is a shaft with discrete thickness grading constructed from eight plies $(\pm\theta \pm\theta)_s$ with the same properties of carbon/epoxy composites (see **Table 2**) and same thicknesses. This sequence is applied in filament wound circular shells, as such a process demands adjacent $(\pm\theta)$ layers. **Figure 9** depicts the obtained contours in the $(V_{f1}\text{-}\theta)$ design space, which are, as seen, monotonic and symmetric about the zero ply angle. A local maximum of $\hat{T}_{cr}$ can be observed near the design point $(V_{f1},\theta) = (0.7, 90^o)$ with $\hat{T}_{cr} = 1$. This figure illustrates that the maximum critical buckling torque can be achieved when the fiber orientation angle is close to 90°. Other case studies including both discrete and continuous grading with several optimal solutions can be found in Ref. [31]. At the start of the optimization process in each case, the shaft wall was divided into a large number of layers with equal thicknesses, for example, $N_L = 32$. It has been found that the optimization algorithm treats the number of layers as an additional implicit variable. Sometimes the computer discards one or more layers by letting their thicknesses sink to the lower limits and sometimes makes some consecutive layers identical, that is, having the same fiber orientation and volume fraction. Such a situation was repeated for many cases of study. It was found that the appropriate number to be taken for the shaft problem under consideration is $N_L = 8$. This would eliminate much of the numerical effort necessary for performing structural analysis in each optimization cycle and, consequently, reduces the computational time considerably.

The final attained optimal solution was a cross ply layup $[90^0/0^0]_4$ with the fiber volume fraction in the eight layers reached its upper value of 70%. The optimal dimensionless ply

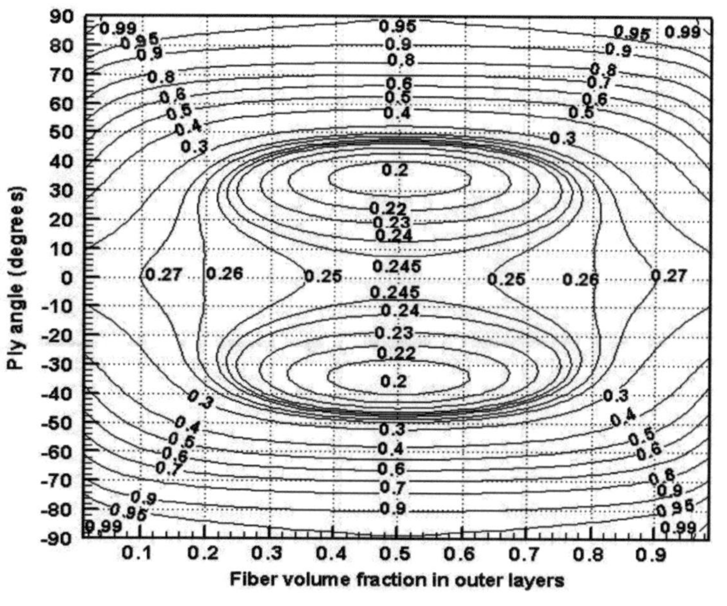

Figure 9. $\hat{T}_{cr}$—contours in $(V_{f1}\text{-}\theta)$ design space under mass constraint $\hat{M} = 1$. (Case of drive shaft with eight symmetric, balanced, carbon/ epoxy layers)

thickness was found to be $[0.1994, 0.0967, 0.152, 0.019]_s$ at which the shaft torsional buckling capacity was increased by *32.1%* above that of the baseline design. However, the total structural mass has reached its baseline value, and whirling constraint became active at the achieved optimum design point.

4.2. Whirling optimization problem

The calculation of the critical speed, also referred to as whirl instability of a rotating shaft, is based on the work given in Refs. [32, 33]. The critical speed is defined as the point at which the spinning shaft reaches its first natural frequency. The shaft is modeled as a Timoshenko beam, which implies that first-order shear deformation theory with rotatory inertia and gyroscopic action was used. The shaft is assumed to be pinned at both ends with a Cartesian coordinate system (x, y, z), where x is measured along the longitudinal axis of the shaft. The displacements in the y and z directions are denoted by v and w, respectively, and Φ is the angle of twist. The cross-sectional area, second moment of area, and polar moment of area are denoted by A, I, and J, respectively. The equations of motion are derived by invoking Hamilton's principle, with the following results [32]:

$$C_B\frac{\partial^2 v}{\partial X^4}+\rho A\frac{\partial^2 v}{\partial t^2}-\left(\rho I+\frac{C_B}{C_S}\rho A\right)\frac{\partial^4 v}{\partial X^2\partial t^2}-2\rho I\Omega\left(\frac{\partial^3 w}{\partial x^2\partial t}-\frac{\rho A}{C_S}\frac{\partial^3 w}{\partial t^3}\right)+\frac{\rho I\rho A}{C_S}\frac{\partial^4 v}{\partial t^4}+(C_{BT}/2)\frac{\partial^3\phi}{\partial x^3}=0 \tag{23}$$

$$C_B\frac{\partial^4 w}{\partial x^4}+\rho A\frac{\partial^2 w}{\partial t^2}-\left(\rho I+\frac{C_B}{C_S}\rho A\right)\frac{\partial^4 w}{\partial x^2\partial t^2}+2\rho I\Omega\left(\frac{\partial^3 v}{\partial x^2\partial t}-\frac{\rho A}{C_S}\frac{\partial^3 v}{\partial t^3}\right)+\frac{\rho I\rho A}{C_S}\frac{\partial^4 w}{\partial t^4}+(C_{BT}/2)\frac{\partial^3\phi}{\partial x^3}=0 \tag{24}$$

$$(C_{BT}/2)\left(\frac{\partial^3 v}{\partial x^3}+\frac{\partial^3 w}{\partial x^3}\right)-(C_{BT}/2)\frac{\rho A}{C_S}\left(\frac{\partial^3 v}{\partial x\partial t^2}+\frac{\partial^3 w}{\partial x\partial t^2}\right)+C_T\frac{\partial^2\phi}{\partial x^2}-\rho J\frac{\partial^2\phi}{\partial t^2}=0 \tag{25}$$

The symbol t denotes time and Ω the rotational speed. The ρA, ρI, and ρJ terms account for translational, rotary, and torsional inertias, respectively, while the $2\rho I$ terms account for the gyroscopic inertia effects. It is assumed that the flexural and bending-twisting coupling rigidities (C_B and C_{BT}) associated with bending about the y and z axes are identical; likewise for the transverse shear stiffness (C_s) [32]. Bert and Kim [33] considered the case of simply supported shaft and assumed separable solution in space and time to solve the associated eigenvalue problem. The derived frequency equation is given by:

$$\begin{gathered}(C_{11}^2-C_{12}^2)C_{33}-2C_{11}C_{13}C_{31}=0\\ C_{11}=C_B\lambda^4-\left(\rho I+\frac{C_B}{C_S}\rho A\right)\omega^2\lambda^2+\left(\frac{\rho I\rho A}{C_S}\omega^2-\rho A\omega^2\right)\end{gathered} \tag{26}$$

$$\begin{gathered}C_{12}=2\rho I\Omega\omega\left(\lambda^2-\frac{\rho A}{C_S}\omega^2\right);C_{13}=(C_{BT}/2)\lambda^3\\ C_{31}=C_{13}-(C_{BT}/2)(\rho A/C_S)\lambda\omega^2;\quad C_{33}=C_T\lambda^2-\rho J\omega^2\end{gathered} \tag{27}$$

where $\lambda = n\pi/L$, ω = circular natural frequency, and n = mode number. For each natural frequency of the nonrotating shaft, the rotational speed (Ω) develops gyroscopic moments, which cause the natural frequency to bifurcate into two. The higher of the two increases with Ω and is associated with forward precession, while the lower one decreases with Ω and is associated with backward precession. A critical instability occurs when the rotational speed coincides with the first backward-precision natural frequency, which is termed as the first critical speed. Two alternatives may be considered regarding the whirling optimization problem [34]:

(a) Direct maximization of the critical rotational speed

Find the design variables vector $\vec{X} = \left(V_f, \theta, \hat{h}\right)_{k=1,2,\ldots N_L}$, which minimizes the objective function:

$$\begin{array}{ll} \text{Minimize} & F = -\hat{\Omega}_{cr} \\ \text{Subject to} & \hat{M} - 1 \leq 0 \end{array} \tag{28}$$

$$\begin{array}{l} \left(\dfrac{\tau_{max}}{\tau_{allow}}\right) - 1.0 \leq 0 \\ \left(\dfrac{T_{max}}{T_{cr,o}}\right) - \hat{T}_{cr} \leq 0 \end{array} \tag{29}$$

(b) Placement of the critical speed

The other alternative of the objective function is defined by:

$$\text{Minimize} \quad F = \left(\hat{\Omega}_{cr} - \hat{\Omega}^*\right)^2 \tag{30}$$

The same set of constraints given in Eq. (23) is applied. The notation $\hat{\Omega}^*$ is a dimensionless target rotational speed, which should be greater than the maximum permissible rotational speed by a reasonable margin (e.g., *10–20%*). As a case study, a drive shaft with continuous material grading along the shaft axis is optimized considering the following power-law model:

$$\boldsymbol{V_f(\hat{x}) = V_f(0.5) + \left[V_f(0) - V_f(0.5)\right]((1 - 2|\hat{x}|)^n)^P}, \tag{31}$$

where $V_f\,(0)$ is the fiber volume fraction at the right or left end of the drive shaft, while $V_f\,(0.5)$ is the fiber volume fraction at the middle of the shaft length. **Figure 10** illustrates the level curves of the normalized critical speed augmented with the mass equality constraint. It is seen that there are four distinct zones separated by the contour lines $\hat{\boldsymbol{\Omega}}_{cr} = 1.0$. The upper left zone and lower right zone contain local maximum solutions. The best point (p, Δ_f) = (4.53, 0.3), corresponding to $\hat{\boldsymbol{\Omega}}_{cr} = 1.045$, is located inside the zone where the fiber taper ratio $\Delta f = V_f(0)/V_f(0.5)$ is less than one. The upper empty zone contains infeasible solutions that violate the imposed constraints. Another case study considers the through-thickness grading pattern

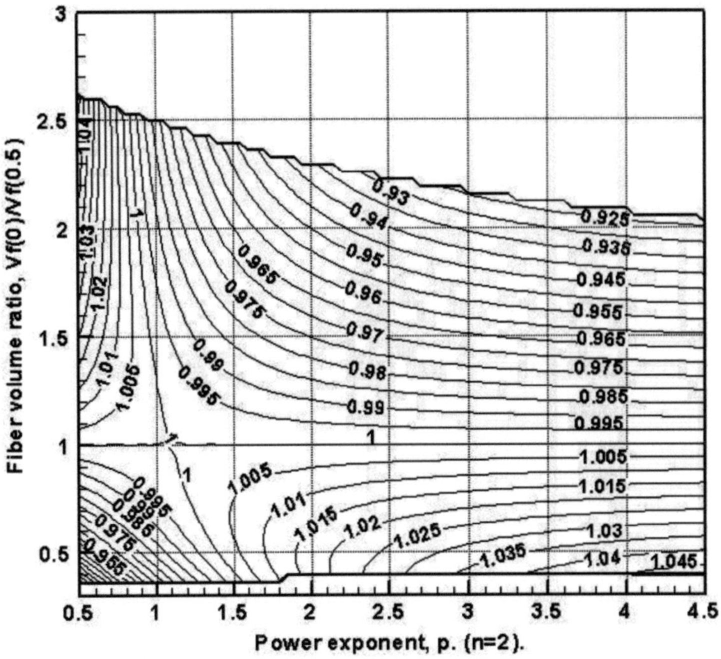

Figure 10. Normalized critical speed $\hat{\Omega}_{cr}$ augmented with the mass constraint ($\hat{M} = 1.0$) in *(p-Δf)* design space.

given by Eq. (3). The corresponding design variable vector is defined by $\vec{X} = \left(V_f(0), V_f\left(\frac{1}{2}\right), p, \hat{H}\right)$ with lower and upper limits $\vec{X_L}$ = (0.3, 0.3, 0, 0.75) and $\vec{X_U}$ = (0.7, 0.7, ∞, 1.25). The attained optimal design variable vector was calculated to be $\vec{X_{opt}} = (0.7, 0.3, 5.61, 0.955)$ at which the maximum critical speed increased by *14%* above that of the baseline design with active mass constraint.

A last optimization strategy to be addressed here is to combine the two criteria in a single objective function subject to the mass, strength, and side constraints.

$$\begin{aligned} &\text{Minimize} \quad F = -\left(\hat{\Omega}_{cr} + \hat{T}_{cr}\right) \\ &\text{Subject to} \quad \hat{M} - 1 \le 0 \\ &\qquad \left(\frac{\tau_{max}}{\tau_{allow}}\right) - 1.0 \le 0 \end{aligned} \tag{32}$$

Eq. (22) assumes that whirling and torsional buckling instabilities are of equal relative importance. This model resulted in a balanced improvement in both stabilities with active mass constraint. The attained optimal solution was found to have a uniform distribution of the fiber volume fraction with its upper limiting value of 70% and wall thickness = 0.935. The corresponding optimal values of the design objectives were $\hat{\Omega}_{cr} = 1.135$ and $\hat{T}_{cr} = 1.161$, representing optimization gains *13.5* and *16.1%*, respectively, as measured from the baseline design.

5. Optimization of FGM wings against divergence

The use of the in-plane grading in aeroelastic design was first exploited by Librescu and Maalawi [6], who introduced the underlying concepts of using material grading in optimizing subsonic rectangular wings against torsional instability. Exact mathematical models were developed allowing the material physical and mechanical properties to change in the wing spanwise direction, where both continuous and piecewise structural models were successfully implemented. In this section, analytical solutions are developed for slender tapered composite wings through optimal grading of the material volume fraction in the spanwise direction. The enhancement of the wing torsional stability is measured by maximization of the critical flight speed at which aeroelastic divergence occurs. The total structural mass is maintained at a value equals to that of a known baseline design in order not to violate other performance requirements. **Figure 11** depicts a slender wing constructed from *Np* panels with trapezoidal planform and known airfoil cross section. The wing is considered to be made of unidirectional fiber-reinforced composites with variable fiber volume fraction in the spanwise direction. The flow is taken to be steady and incompressible, and the aspect ratio is assumed to be sufficiently large so that the classical engineering theory of torsion can be applicable and the state of deformation described in terms of one space coordinate.

The chord distribution is assumed to have the form:

$$C(x) = C_r\ (1 - \beta_c x), \beta_c = (1 - \Delta_c) \tag{33}$$

The symbol Δ_c denotes the chord taper ratio (= tip chord C_t/root chord C_r) and x (= x_1/L) denotes the dimensionless spanwise coordinate. The equivalent shear modulus G of a unidirectional reinforced composite, thin-walled cross section can be determined from the relation [35]:

$$G = f_1\ G_{12} \tag{34}$$

where f_1 is a function that depends on the geometry and thickness ratio of the cross section (h/C) and the ratio $(G_{12}/\ G_{13})$, where G_{12} and G_{13} are the in-plane and out-of-plane shear moduli,

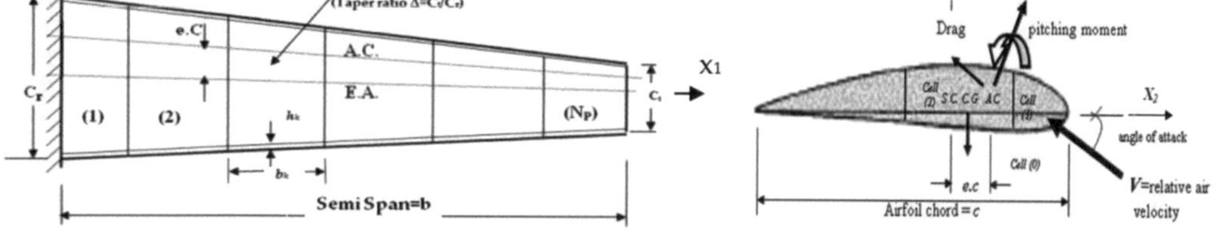

(a) (b)

Figure 11. Trapezoidal wing planform and cross section geometry. (a) Multipanel, piecewise wing model, (b) airfoil section and applied airloads.

respectively (refer to **Table 1**). C is the chord length and h is the maximum thickness of the cross section. For many types of fibrous composites that are commonly utilized in aerospace industry [23], such as carbon/epoxy and graphite epoxy, both moduli are approximately equal, $G_{12} \approx G_{13}$.

Using the classical elasticity and aerodynamic strip theories, the governing differential equation of torsional stability in dimensionless form is [35]:

$$\left(\hat{G}\hat{J}\alpha'\right)' + \hat{V}^2\alpha(x) = 0 \tag{35}$$

The associated boundary conditions of the elastic angle of attack, α, are $\alpha(0) = 0$ and $\alpha'(1) = 0$. The symbol $\hat{G} = G_{12}/G_{12,o}$ denotes the dimensionless shear modulus, $\hat{J} = J/J_r$ denotes the dimensionless torsion constant, and the prime denotes differentiation with respect to the dimensionless coordinate $x = x_1/L$. The dimensionless flight speed is defined by $\hat{V} = VC_r b\sqrt{\rho a e/2GJ_r}$, where $(GJ)_r$ is the torsional stiffness of the baseline design at root. The shear modulus $G_{12,0}$ of the baseline design can be calculated by taking $V_{fo} = 50\%$.

Considering the K-th panel of the wing as shown in **Figure 11a,** and using the transformation $y = (1\text{-}\beta x)$, Eq. (25) takes the form:

$$z\alpha'' + 3\alpha' + a_k^2\alpha = 0, \quad (1\text{-}\beta x_{k+1}) \le y \le (1 - \beta x_k) \tag{36}$$

where $a_k = \hat{V}/\beta\sqrt{\hat{h}_k\hat{G}_k}$, $\hat{h}_k$ and $\hat{G}_k$ are the normalized wall thickness and shear modulus of the kth wing panel, respectively. The general solution of Eq. (26) is:

$$\alpha(y) = A_1\frac{J_2\left(2\sqrt{a_k y}\right)}{y} - A_2\frac{Y_2\left(2\sqrt{a_k y}\right)}{y} \tag{37}$$

where J_2 and Y_2 are Bessel's function of the first and second kind with order 2, respectively [35], and A_1 and A_2 are the constants of integration. The dimensionless internal torsional moment, T, can be obtained by differentiating Eq. (27) and multiplying by the dimensionless shear rigidity. Applying the boundary conditions at stations (k) and $(k + 1)$, the constants A_1 and A_2 can be expressed in terms of the state variables at station (k), which can be related to those at station $(k + 1)$ by the transfer matrix relation:

$$\begin{Bmatrix} \alpha_{k+1} \\ T_{k+1} \end{Bmatrix} = \left[E^{(k)}\right]\begin{Bmatrix} \alpha_k \\ T_k \end{Bmatrix} \tag{38}$$

It is now possible to compute the state variables progressively along the wing span by applying continuity requirements of the variables (α, T) among the interconnecting boundaries of the various wing panels. The divergence speed can be calculated by applying the boundary conditions and considering the nontrivial solution of the resulting equations (similar to the procedure outlined in Section 3.1.3). The associated optimization problem may be cast in the following:

$$
\begin{aligned}
&\text{Minimize} - \hat{V}_{div} \\
\text{Subject to} \quad &\hat{M}_s - 1.0 \le 0 \\
&(0.25, 0.5, 0.0) \le \left(V_f, \hat{h}, \hat{b}\right)_{k=1,2,..Np} \le (0.75, 1.25, 1.0) \\
&\sum \hat{b}_k = 1.0
\end{aligned}
\tag{39}
$$

The preassigned parameters that do not change during the optimization process include the wing semispan (b), the chord taper ration (Δ_c), airfoil type and geometry, and fiber and resin material types. This model has been applied to obtain wing designs with improved torsional stability by maximizing the divergence speed (V_{div}) without weight penalty. The selected material is carbon-*AS4*/epoxy-*3501-6* composite (see **Table 2**), which has favorable characteristics and is highly desirable in manufacturing aircraft structures. The baseline design has uniform mass and stiffness distributions and is made of uniform unidirectional fibrous composite with equal volume fraction of the matrix and fiber materials, that is, V_{fo} = *50%*. **Figures 12** and **13** show the developed level curves of constant divergence speed (also named isodiverts) for two-panel wings with chord tapering ratio, Δ_c = *0.5*. Actually, these curves represent the dimensionless critical speed, augmented with the equality mass constraint. Examining **Figure 12**, it is seen that the V_{div} function is well behaved and continuous everywhere in the selected design space except in the empty regions to the upper left and right regions, where the equality mass constraint is violated. The feasible domain is bounded from

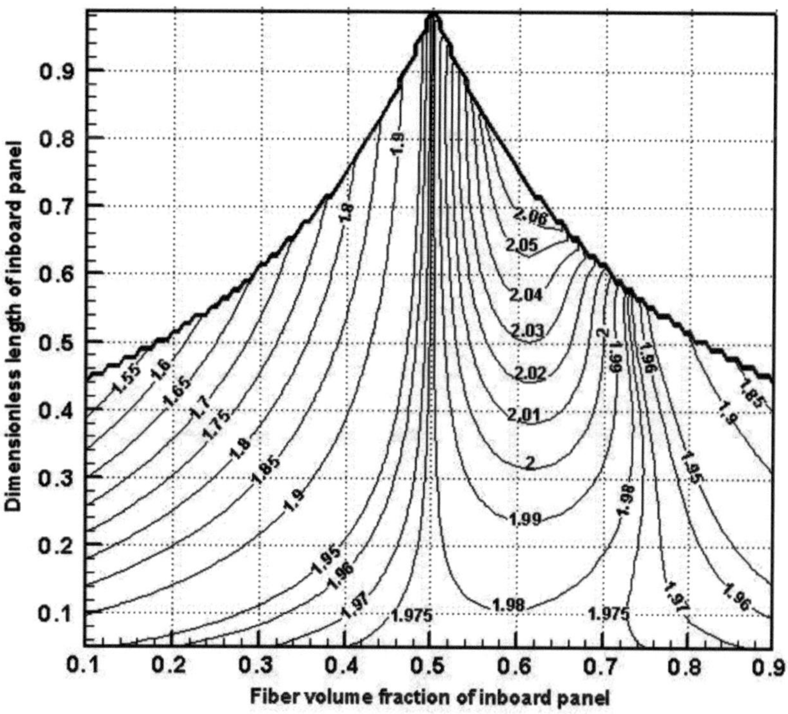

Figure 12. Isodivert in (V_{f1}- b_1) design space for a two-panel wing model ($h_1 = h_2 = 1.0$, $\Delta_c = 0.5$, $\hat{M}_s = 1$).

above by the two curved lines representing the upper and lower limiting constraints imposed on the volume fraction of the outboard blade panel. The contours inside the feasible domain are not allowed to penetrate these borderlines and obliged to turn sharply to be asymptotes to them, in order not to violate the mass constraint. The final attained optimal solutions are summarized in **Table 4**. It can be observed that good wing patterns shall have the lower limit of the fiber volume fraction at the tip and the upper limit at root. Using material and wall thickness grading together results in a considerable enhancement of the wing torsional stability.

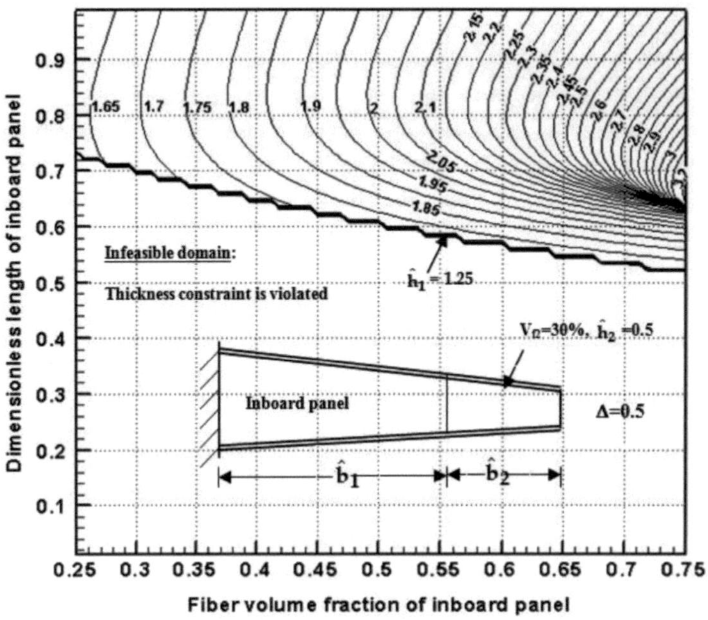

Figure 13. Isodivert in (V_{f1}- b_1) design space for a two-panel wing model (h_2 = 0.5, V_{f2} = 0.3, Δ_c = 0.5, $\hat{M}_s = 1$).

N_p	Type of grading	$\vec{X}_{opt} = \left(V_f, \hat{h}, \hat{b}\right)_{k=1,2,..Np}$	$\hat{V}_{div,max}$	*Optimization gain %*
2	Material	$(0.5906, 1.0, 0.6594)_1$ $(0.2500, 1.0, 0.3406)_2$	2.0523	3.93%
	Material and thickness	$(0.75, 1.0625, 0.7)_1$ $(0.25, 0.5250, 0.3)_2$	3.2275	63.44%
3	Material	$(0.6000, 1.0, 0.5750)_1$ $(0.4125, 1.0, 0.1375)_2$ $(0.2500, 1.0, 0.2875)_3$	2.05625	4.13%
	Material and thickness	$(0.75, 1.10, 0.65)_1$ $(0.30, 0.65, 0.05)_2$ $(0.25, 0.50, 0.3)_3$	3.3475	69.52%

Table 4. Optimal piecewise wing designs with chord taper = *0.5* (baseline value = *1.9747*).

http://dx.doi.org/10.5772/intechopen.82411

6. Optimization of composite thin-walled pipes conveying fluid

The subject of vibration and stability of thin pipes conveying flowing fluids is of a considerable practical interest. An advanced textbook by Païdoussis [36] gives an excellent review of the several developments made in this research area. Practical models for enhancing static and dynamic stability characteristics of pipelines constructed from uniform modules were addressed by Maalawi et al. [37, 38], where the relevant design variables were selected to be the mean diameter, wall thickness, and length of each module composing the pipeline. The general case of an elastically supported pipe, covering a variety of boundary conditions, was also investigated. Distinct domains of the flutter instability boundaries were presented for different ratios of the fluid-to-pipe mass, and the variation of the critical flow velocity with support flexibility was examined and discussed. Concerning pipelines made of advanced *FGMs*, this section presents a mathematical model for enhancing the overall system stability against flutter and/or divergence under mass constraint. **Figure 14** shows a *FGM* pipe conveying flowing fluid with the coordinate system chosen such that the x-axis coincides with the longitudinal centroidal axis in its undeformed position, while the y- and z-axes coincide with the cross section principal axes. The pipe model consists of rigidly connected thin-walled circular tubes made of unidirectional fibrous composite material. Each pipe module has different material properties, wall thickness, and length. Such a configuration results in a piecewise axial grading of either the material of construction or the wall thickness in the direction of the pipe axis. Assuming no voids are present, the distributions of the mass density (ρ) and modulus of elasticity (E) can be determined using the formulas of **Table 1**.

The associated eigenvalue problem is described by the fourth-order ordinary differential equation [39]:

$$(EI)_k V'''' + U^2 V'' + 2i\omega U\sqrt{\beta_o} V' - m_k \omega^2 V = 0 \tag{40}$$

where *V(x)* is the dimensionless mode shape satisfying boundary conditions, and ω is the corresponding dimensionless frequency of oscillation, which will be, in general, a complex number to be determined by the requirement of nontrivial solutions, $V(x) \neq 0$. More details for

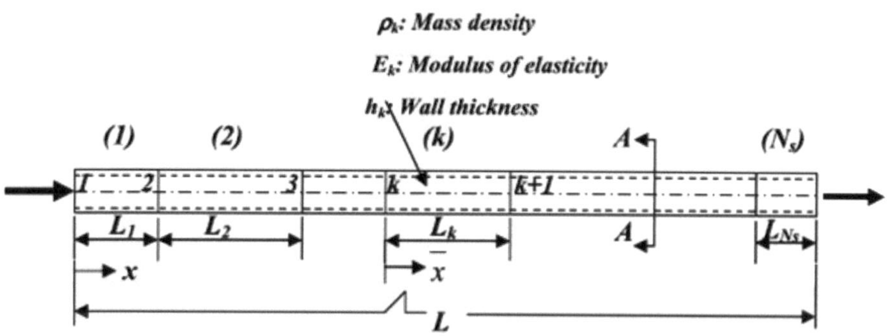

Figure 14. Multimodule composite pipe conveying flowing fluid.

the definition of the various parameters and dimensionless design variables are given in Ref. [39]. Both static and dynamic instability phenomena can be involved for the physical model described by Eq. (30), depending on the type of boundary conditions at the pipe ends. The notation ()′ means total spatial derivative. The general solution can be obtained using standard power series methods:

$$V(x) = \sum_{j=1}^{4} A_j e^{ip_j x} \tag{41}$$

p_j, $j = 1,\ldots 4$ are the four roots of the fourth-order polynomial:

$$p^4 - (\alpha_k U^2)p^2 - \left(2\omega U\sqrt{\beta_o}\alpha_k\right)p - m_k\omega^2\alpha_k = 0 \quad , \alpha_k = 1/(E_k h_k) \tag{42}$$

An efficient method [38] is successfully implemented to find the complex roots of Eq. (32) by formulating a special companion matrix and finding the associated eigenvalues for any desired values of the variables α_k, β_o, ω, and U. The transfer matrix $[T_k]$ of the kth pipe module can be obtained by performing the matrix multiplications

$$[T_k] = \left[P^{(k)}\right]\left[E^{(k)}\right]\left[P^{(k)}\right]^{-1} \tag{43}$$

$$\left[p^{(k)}\right] = \begin{bmatrix} 1 & 1 & 1 & 1 \\ -ip_1 & -ip_2 & -ip_3 & -ip_4 \\ (Eh)_k p_1^2 & (Eh)_k p_2^2 & (Eh)_k p_3^2 & (Eh)_k p_4^2 \\ i(Eh)_k p_1^3 & i(Eh)_k p_2^3 & i(Eh)_k p_3^3 & i(Eh)_k p_4^3 \end{bmatrix} \tag{44}$$

and $[E]$ is a diagonal matrix with elements:

$$E_{jj}^{(k)} = e^{ip_j^{(k)} L_k} \tag{45}$$

Applying the appropriate boundary conditions and considering only the nontrivial solution, the resulting characteristic equation can be solved numerically for the frequency and the critical flow velocity. The system is stable or unstable according to whether the imaginary component of the frequency, ω, is positive or negative, respectively. In case of neutral stability, ω is wholly real. As the flow velocity increases, the system may become unstable in one or more of its normal modes. The critical flow velocity is the greatest velocity for which the system is stable in all its modes. The characteristic equation for cantilevered pipe is:

$$T_{33}T_{44} - T_{34}T_{43} = 0, \tag{46}$$

where T_{ij} are the elements of the overall transfer matrix. For a specific mode number, the proper starting frequency is determined by solving a subsidiary eigenvalue problem corresponding to the case of stationary fluid inside the vibrating pipe. A globally convergent optimization algorithm, known as Levenberg-Marquardt algorithm [38], has been applied to solve the resulting nonlinear equations derived from the consideration of nontrivial solution of characteristic equation. The effect of flow for small velocity is to damp the system in all modes. At higher velocities, some of the modes become less damped and the corresponding branches cross the Re(ω)-axis, indicating the existence of unstable oscillations of the system. If a branch passes through the origin, that i, $\omega = 0$, the case of static instability (called divergence) is reached.

6.1. Flutter solutions

To verify the developed formulation, the classical problem dealing with one-module cantilevered pipe is considered first. The dimensionless wall thickness and length of the pipe are assigned at a value of *1.0*, while the volume fraction at *50%*. **Figure 15** depicts variation of the critical flutter velocity and frequency with the mass density ratio *MRo*, covering a wide range of pipe and fluid mass densities. It is seen that there are four subdomains with the associated flutter modes defined in the specified intervals of the mass ratio. The upper and lower bounds determine the critical values of the mass ratios at which some of the frequency branches cross each other at the same value of the flutter velocity. The overall flutter mode may be regarded as composed of different quasimodes separated at the shown "jumps" in the U_f-MR_o curve. The calculated mass density ratios at the three indicated frequency jumps are 0.4225, 2.29, and 12.33, respectively. They correspond to multiple points of neutral stability, where for a finite incremental increase in the flow velocity, the system becomes unstable, then regains stability, and then once again becomes unstable with a noticeable abrupt increase in the flutter frequency. Next, we consider a baseline design made of carbon/epoxy composites (see **Table 2**) with mass ratio $MR_o = 2.0$. The calculated values of the dimensionless flutter velocity and frequency are found to be $U_f = 10.78$ and $\omega_f = 26$, respectively. Keeping the total dimensionless mass constant at the value corresponding to the baseline design, the best solution having the greatest flutter velocity was found to be $(V_f, h_1) = (0.70, 0.9345)$, which corresponds to $U_f = 12.517$ and $\omega_f = 31.8615$.

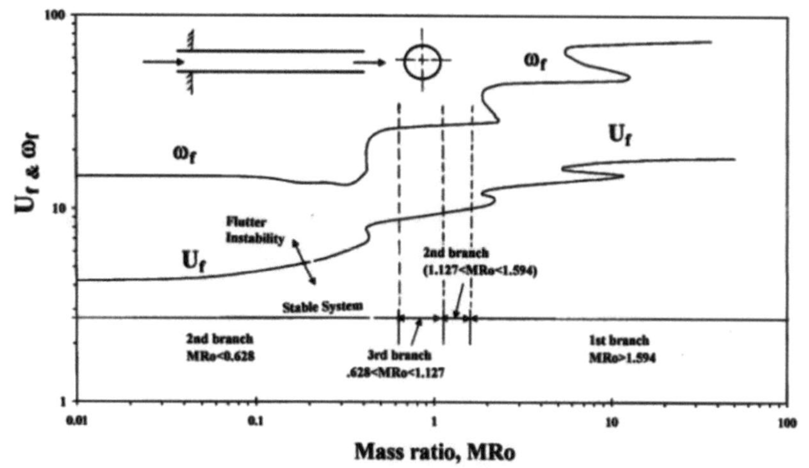

Figure 15. Variation of flutter speed and frequency with mass ratio for a uniform one-module cantilevered pipe ($V_{f1} = 50\%$, $h_1 = 1$, $L_1 = 1$).

6.1.1. Solutions for cantilevered two-module pipe with uniform thickness ($h_1 = h_2 = 1.0$)

Considering the case of two-module pipe, a direct and fast way for checking out system stability for any desired set of the dimensionless design variables $(V_{f1}, L)_{k=1,2}$ is given here. Lower and upper bounds are imposed on the design variables in order not to violate other strength and manufacturing requirements. The fiber volume fraction is constrained to be within the range 30% up to 70%, while the dimensionless length is between 0.0 and 1.0. The mass ratio *MRo* is taken to be 2.0.

Dimensionless flutter velocity and flutter frequency are obtained from the frequency and velocity branches at the four modes. The lowest frequency and velocity among the four modes at which Imag(ω) = 0.0 are considered the flutter velocity and frequency. These computed values at different conditions are employed in constructing the flutter velocity and frequency contours as shown in **Figure 16**. The white regions shown in both figures indicate that the fiber

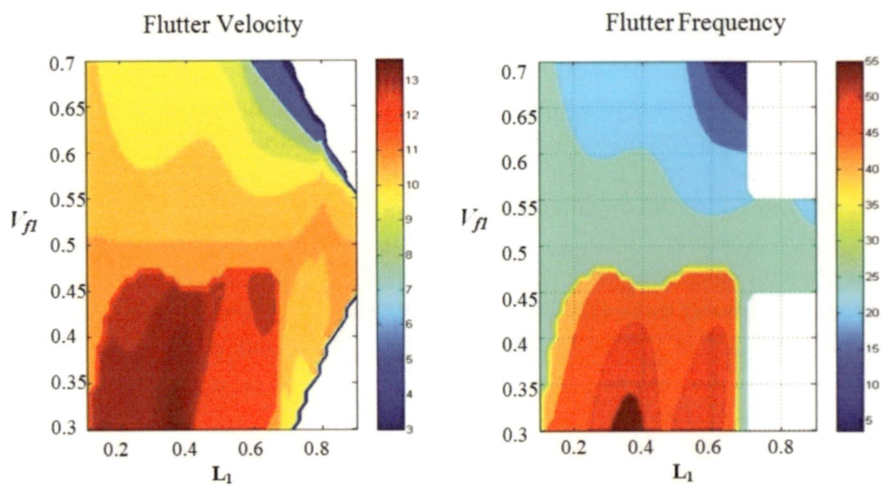

Figure 16. Contour plots of flutter velocity and frequency in (V_{f1}-L_1) design space.

Design variables: $(V_f, L)_{k=1,2}$	Dimensionless $U_{flutter}$	Stability improvement %
(0.30, 0.20), (0.55, 0.80)	13.15	21.99
(0.30, 0.50), (0.70, 0.50)	12.07	11.97
(0.35, 0.25), (0.55, 0.75)	13.04	20.96
(0.35, 0.40), (0.60, 0.60)	13.31	23.47 (max.)
(0.40, 0.50), (0.60, 0.50)	12.70	17.81
(0.40, 0.60), (0.65, 0.40)	12.77	18.46
(0.45, 0.50), (0.55, 0.50)	12.82	18.92
(0.45, 0.75), (0.65, 0.25)	11.73	8.81
(0.50, 0.50), (0.50, 0.50)	10.78	0.00 (baseline)

Table 5. Standard solutions for a cantilevered two-module pipeline with uniform thickness ($h_1 = h_2 = 1.0$, $M = 1.0$).

volume fraction of the material of the second segment does not fall in range between 0.3 and 0.7. The maximum flutter velocity (U_f) and its corresponding flutter frequency (ω_f) occur in the region colored with dark brown $L_1 = 0.36$ and $V_{f1} = 0.3$. The maximum U_f and its corresponding ω_f are 13.67 and 58.4, respectively. **Table 5** gives several standard optimal solutions for the two-module case study. The global optimum design point is seen to be $(V_f, L)_{k=1,2} = (0.35, 0.40), (0.60, 0.60)$, at which the normalized flutter velocity reached a value of 13.31 corresponding to 23.47% optimization gain.

7. Conclusions

As a major concern in producing efficient structures with enhanced properties and tailored response, this chapter presents appropriate design optimization models for improving performance and operational efficiency of different types of composite structural members. The concept of material grading has been successfully applied by incorporating the distribution of the volume fractions of the composite material constituents in the mathematical model formulation. Various scenarios in modeling the spatial variation of material properties of functionally graded structures are addressed. The associated optimization strategies include frequency maximization of thin-walled composite beams, optimization of drive shafts against torsional buckling and whirling instability, and maximization of the critical flight speed of subsonic aircraft wings. Design variables encompass the distribution of volume fraction, ply angle, and wall thickness as well. Detailed optimization models have been formulated and presented for improving the dynamic performance and increasing the overall stiffness-to-mass level of thin-walled composite beams. The objective functions have been measured by maximizing the natural frequencies and place them far away from the excitation frequencies, while maintaining the total structural mass at a constant value. For discrete models, the optimized beams can be constructed from any arbitrary number of uniform segments where the length of each segment has shown to be an important variable in the optimization process. It has also been proved that expressing all parameters in dimensionless forms results in naturally scaled design variables, constraints, and objective functions, which are favored by a variety of optimization algorithms. The attained optimal solutions using continuous grading depend entirely upon the prescribed power-law expression, which represents additional constraint on the optimization problem. Results show that material grading in the spanwise direction is much more better than grading through the wall thickness of the cross section. Regarding optimization of *FGM* drive shafts, it was shown that the best model is to combine torsional buckling and whirling in a single objective function subject to mass constraint. This has produced a balanced improvement in both stabilities with active mass constraint at the attained optimal design point.

In the context of aeroelastic stability of aircraft structures, an analytical model has been formulated to optimize subsonic trapezoidal wings against divergence. It was shown that by using material and thickness grading simultaneously, the aeroelastic stability boundary can be broaden by more than 50% above that of a known baseline design having the same total structural mass. Other stability problems concerning fluid-structure interaction have also been addressed. Both flutter and divergence optimization have been considered, and several design charts that are useful for direct determination of the optimal values of the design variables are given. It has been confirmed that the segment length is the most significant design variable in the whole optimization process. Some investigators who apply finite elements have not recognized that the length of each element can be taken as a main design variable in the whole set of optimization variables. The results from the present approach reveal that piecewise grading of the material can be promising in producing truly efficient structural designs with enhanced stability, dynamic, and aeroelastic performance. It is the author's wish that the results presented in this chapter will be compared and validated through other optimization techniques such as genetic algorithms or any appropriate global optimization algorithm.

Actually, the most economic structural design that will perform its intended function with adequate safety and durability requires much more than the procedures that have been described in this chapter. Further optimization studies must depend on a more accurate analysis of constructional cost. This combined with probability studies of load applications and materials variations should contribute to further efficiency achievement. Much improved and economical designs for the main structural components may be obtained by considering multidisciplinary design optimization, which allows designers to incorporate all relevant design objectives simultaneously. Finally, it is important to mention that, while FGM may serve as an excellent optimization and material tailoring tool, the ability to incorporate optimization techniques and solutions in practical design depend on the capacity to manufacture these materials to required specifications. Conventional techniques are often incapable of adequately addressing this issue. In conclusion, FGMs represent a rapidly developing area of science and engineering with numerous practical applications. The research needs in this area are uniquely numerous and diverse, but FGMs promise significant potential benefits that fully justify the necessary effort.

Author details

Karam Maalawi

Address all correspondence to: maalawi@netscape.net

Mechanical Engineering Department, National Research Centre, Cairo, Egypt

References

[1] Suresh S, Mortensen A. Fundamentals of Functionally Graded Materials. London: IOM Communications Ltd; 1998. 165 p. ISBN: 9781861250636

[2] Miyamoto Y, Kaysser WA, Rabin BH, Kawasaki A, Ford RG. Functionally Graded Materials: Design, Processing and Applications. Springer Science & Business Media; 2013. 330 p. ISBN: 9781461553014

[3] Lipton R. Design of functionally graded composite structures in the presence of stress constraints. International Journal of Solids and Structures. 2002;**39**:2575-2586

[4] Cho JR, Shin SW. Material composition optimization for heat-resisting FGM by artificial neural network. Composites Part A Applied Science and Manufacturing. 2004;**35**:585-594

[5] Goupee AJ, Vel SS. Optimization of natural frequencies of bidirectional functionally graded beams. Journal of Structural and Multidisciplinary Optimization. 2006;**32**(6):73-484

[6] Librescu L, Maalawi K. Material grading for improved aeroelastic stability of composite wings. Journal of Mechanics of Materials and Structures. 2007;**2**(7):1381-1394

[7] Birman V, Byrd WL. Modeling and analysis of functionally graded materials and structures. Applied Mechanics Reviews. 2007;**60**(5):195-216

[8] Elishakoff I, Guede Z. Analytical polynomial solutions for vibrating axially graded beams. Journal of Mechanics and Advanced Materials and Structures. 2004;**11**:517-533

[9] Qian L, Batra R. Design of bidirectional functionally graded plate for optimal natural frequencies. Journal of Sound and Vibration. 2005;**280**:415-424

[10] Maalawi KY. Functionally graded bars with enhanced dynamic performance. Journal of Mechanics of Materials and Structures. 2011;**6**(1-4):377-393

[11] Elishakoff I, Endres J. Extension of Euler's problem to axially graded columns. Journal of Intelligent Material Systems and Structures. 2005;**16**(1):77-83

[12] Li S-R, Batra R. Buckling of axially compressed thin cylindrical shells with functionally graded middle layer. Journal of Thin-Walled Structures. 2006;**44**:1039-1047

[13] Maalawi KY. Optimization of elastic columns using axial grading concept. Engineering Structures. 2009;**31**(12):2922-2929

[14] Maalawi KY. Use of material grading for enhanced buckling design of thin-walled composite rings/long cylinders under external pressure. Composite Structures. 2011;**93**(2):351-359

[15] Shin-Yao K. Flutter of rectangular composite plates with variable fiber spacing. Composite Structures. 2011;**93**:2533-2540

[16] Maalawi KY. Stability dynamic and aeroelastic optimization of functionally graded composite structures. In: Coskun SB, editor. Advances in Computational Stability Analysis. Rijeka: InTech; 2012. pp. 17-42. DOI: 10.5772/45878.ch2

[17] Rao S. Engineering Optimization: Theory and Practice. 4th ed. New York: John Wiley & Sons; 2009. ISBN: 978-0470183526

[18] Maalawi K, Badr M. Design optimization of mechanical elements and structures: A review with application. Journal of Applied Sciences Research. 2009;**5**(2):221-231

[19] Venkataraman P. Applied Optimization with MATLAB Programming. 2nd ed. New York: John Wiley & Sons; 2009. ISBN: 978-0470084885

[20] Chen WH, Gibson RF. Property distribution determination of non-uniform composite beams from vibration response measurements and Galerkin's method. Journal of Applied Mechanics. 1998;**65**:127-133

[21] Chi SH, Chung YL. Mechanical behavior of functionally graded material plates under transverse load, I: Analysis. International Journal of Solids and Structures. 2006;**43**: 3657-3674

[22] Bedjilili Y, Tounsi A, Berrabah HM, Mechab I. Natural frequencies of composite beams with a variable fiber volume fraction including rotary inertia and shear deformation. Applied Mathematics and Mechanics. 2009;**30**(6):717-726

[23] Daniel IM, Ishai O. Engineering Mechanics of Composite Materials. 2nd ed. New York: Oxford University Press; 2006

[24] Whitney JM. Elastic moduli of unidirectional composites with anisotropic filaments. Journal of Composite Materials. 1967;**1**:188-194

[25] Maalawi KY. Dynamic optimization of functionally graded thin-walled box beams. International Journal of Structural Stability and Dynamics. 2017;**17**(9):1-24

[26] Dancila DS, Armanios EA. The influence of coupling on the free vibration of anisotropic thin-walled closed-section beams. International Journal of Solids and Structures. 1998; **35**(23):3105-3119

[27] Kieback B, Neubrand A, Riedal H. Processing techniques for functionally graded materials. Materials Science and Engineering. 2003;**362**:81-105

[28] Maalawi KY, El-Chazly NM. Global optimization of multi-element beam-type structures. In: Proceedings of the Second International Conference on Advances in Structural Engineering and Mechanics (ASEM'02). Busan, South Korea; August 2002. pp. 21-23

[29] Bijagare Anup A, Mehar PG, Mujbaile VN. Design optimization & analysis of drive shaft. VSRD International Journal of Mechanical, Automobile & Production Engineering. 2012; **2**(6):210-215

[30] Bert CW, Chun-Do K. Analysis of buckling of hollow laminated composite drive shafts. Composites Science and Technology. 1995;**53**:343-351

[31] Nasr MF, El-Zoghby AA, Maalawi KY, Azzam BS, Badr MA. Torsional buckling optimization of composite drive shafts. World Applied Sciences Journal. 2015;**33**(3):517-524

[32] Bauchau OA. Optimal design of high speed rotating graphite/epoxy shafts. Journal of Composite Materials. 1983;**17**(3):170-180

[33] Bert CW, Chun-Do K. Whirling of composite-material drive shafts including bending-twisting coupling and transverse shear deformation. Journal of Vibration and Acoustics, ASME. 1995;**117**:17-21

[34] Nasr MF, Maalawi KY, Badr MA. Optimization of composite drive shafts against whirling instability. In: Proceedings of the 8th International Conference on Mechanical and Aerospace Engineering (ICMAE); 22-25 July 2017; Prague. Czech Republic: IEEE; 2017. pp. 816-821. DOI: 10.1109/ICMAE.2017.8038756

[35] Librescu L, Maalawi KY. Aeroelastic design optimization of thin-walled subsonic wings against divergence. Thin-Walled Structures. 2009;**47**:89-97

[36] Païdoussis MP. Fluid Structure Interactions: Volume I; Slender Structures and Axial Flow. 2nd ed. New York: Academic Press; 2013. 880 p. ISBN: 9780123973122

[37] Maalawi KY, Ziada MA. On the static instability of flexible pipes conveying fluid. Journal of Fluids and Structures. 2002;**16**(5):685-690

[38] Maalawi KY. Abouel Fotouh AM: Investigation of instability of flexible pipes transporting fluids. Journal of Engineering and Applied Science. 2005;**52**(6):1199-1217

[39] Maalawi KY, Abouel-Fotouh AM, El Bayoumi M, Ali Yehia KA: Design of composite pipes conveying fluid for improved stability characteristics. International Journal of Applied Engineering Research. 2016;**11**(12):7633-7639. ISSN 0973-4562